T0065358

CONFESSIONS

OF AN

ECO-
WARRIOR

Also by Dave Foreman

Ecodefense
The Big Outside

CONFESSIONS
OF AN
ECO-
WARRIOR

DAVE FOREMAN

CROWN TRADE PAPERBACKS / NEW YORK

To
David Brower
and Celia Hunter
with thanks for showing me
the promise of our species

Published by Crown Publishers, Inc.,
201 East 50th Street, New York, New York 10022.
Member of the Crown Publishing Group.
Random House, Inc. New York, Toronto, London, Sydney, Auckland

CROWN TRADE PAPERBACKS and colophon are trademarks of Crown Publishers, Inc.

Manufactured in the United States of America

Library of Congress Cataloging-in-Publication Data
Foreman, Dave, 1946-
    Confessions of an eco-warrior / David Foreman.
        p. cm.
        1. Environmental policy—United States. 2. Environmental policy-United States—Citizen participation. I. Title.
HC110.E5F64 1991
363.7'0574'092—dc20
[B]
                                        90-48729
                                            CIP

ISBN 0-517-88058-0

Contents

CONTENTS

Introduction

Let me be honest about a few things at the very beginning. First of all, I'm not sure what this book is. It's not an autobiography or even a memoir. It's not a work of conservation history or a technical discussion about environmental issues. It's not a comprehensive exposition of biocentric philosophy or an in-depth discussion of conservation biology. It's not a polemic or a how-to guide. I guess it's a little bit like an ugly mongrel dog in which you can see the ears of one breed, the jowls of another, and so on. Whatever this poor bastard is, I hope it gets you, the reader, thinking or wondering, whether you like it or not.

My second caveat is like the warnings on the edges of medieval maps: "Here there be monsters." The monsters in this case are inconsistencies. Ralph Waldo Emerson, who knew what he was talking about, said that a foolish consistency is the hobgoblin of little minds. There are hobgoblins in my mind—and it's not necessarily big—but foolish consistency is one that doesn't stumble around inside it very often.

I must also admit that after twenty years as a full-time conservationist, I have more questions than answers. This is not an operating manual for spaceship Earth or a brilliant solution to the problems of our time. My environmental career is not a linear progression from moderate to radical. It's more of a spiral, like Aldo Leopold's round river flowing back into itself around and around. In the 1970s, when I worked for The Wilderness Society, I believed that being moderate and reasonable was the way to protect the greatest amount of wilderness. I disavowed that in 1980 with the founding of Earth First!, arguing that we should let our actions set the finer points of our philosophy. As you will read, I am not so sure of that approach today. I do not write during a static time. Today is a period of transition—for me personally, for the Earth First! movement, for which I have spoken these last ten years, and for human civilization. In recognition of that, the arrangement of chapters follows a rough chronology from the formation of the Earth First! movement in 1980 to my leaving Earth First! today. As a result, one may perceive a shift in my point of view and in what I advocate. The sequence also goes from describing who I am and what Earth First! is, to the importance of wilderness, to monkeywrenching, to proposals for the future.

The last warning is that I am no saint. We are often lectured to clean up our own acts before poking our noses into other people's business. Well, there isn't time for me to achieve perfection before trying to save the Earth. It's got to be done now. A doctor or nurse working in an emergency room doesn't have time to worry about her own cholesterol while she's trying to bring back a heart attack victim. This is a book about the destruction of Earth by someone who is part of that problem. We are all part of the problem.

What I hope this book does, then, is to motivate people into action. We live in perilous times. The peril is of our own making, and many of us probably deserve it. But the children, and the native peoples of this world, and, most important, all the other species sashaying around in this great dance of life don't deserve the peril we have created.

The ecologist Raymond Dasmann says that World War III has already begun, and that it is the war of industrial humans against the

Earth. He is correct. All of us are warriors on one side or another in this war; there are no sidelines, there are no civilians. Ours is the last generation that will have the choice of wilderness, clean air, abundant wildlife, and expansive forests. The crisis *is* that severe.

If there is one theme that consistently runs through this book, other than the intrinsic value of all natural things and the need for personal action by every one of us, it is an embracing of diversity. I am no moral relativist. I believe that some things are good and some are bad. I have a value system. But I passionately believe that there are many ways to do good, countless methods to defend things of value. My way is not the only good way. My style is not the only valid style.

Thus I do not want to tell anyone what to do. Each of us has to find his or her own role, style, and tools to use in defense of Earth. These may change through time and situation. Wearing a suit and tie, I have sat in a room alone with a United States senator, going over boundary lines on a map of a Wilderness proposal. I've also worn camouflage, prowling around a forest, pulling up survey stakes for a road. What is important is that you do something. Now.

1 | In Time of Crisis

In wildness is the preservation of the world.
—Henry David Thoreau

We are living now in the most critical moment in the three-and-a-half-billion-year history of life on Earth. For this unimaginably long time, life has been developing, expanding, blossoming, and diversifying, filling every available niche with different manifestations of itself, intertwined in complex, globe-girdling relationships. But today this diversity of perhaps 30 million species faces radical and unprecedented change. Never before—not even during the mass extinctions of the dinosaurs at the end of the Cretaceous era, 65 million years ago—has there been such a high rate of extinction as we are now witnessing, such a drastic reduction in the planet's biological diversity.

Over the last three or four hundred years, human civilization has declared war on large mammals, leading some respected ecologists to assert that the only large mammals living twenty years from now will be those we humans choose to allow to live. Other prominent biologists, looking aghast on the wholesale devastation of tropical rain forests and temperate-zone old-growth forests, rapidly accelerating desertification, rapacious commercial fishing, and wasting

1

of high-profile large mammals like whales, elephants, and Tigers ("charismatic megafauna") owing to habitat destruction and poaching, say that Earth could lose one-quarter to one-third of *all* species within forty years.

Not only is this blitzkrieg against the natural world destroying ecosystems and their associated species, but our activities are now beginning to have fundamental, systemic effects upon the entire life-support apparatus of the planet: upsetting the world's climate; poisoning the oceans; destroying the atmospheric ozone layer that protects us from excessive ultraviolet radiation; changing the CO_2 ratio in the atmosphere and causing the "greenhouse effect"; and spreading acid rain, radioactive fallout, pesticides, and industrial contamination throughout the biosphere. Indeed, Professor Michael Soulé, founder of the Society for Conservation Biology, recently warned that vertebrate evolution may be at an end due to the activities of industrial humans.

Clearly, in such a time of crisis, the conservation battle is not one of merely protecting outdoor recreation opportunities, or a matter of aesthetics, or "wise management and use" of natural resources. It is a battle for life itself, for the continued flow of evolution. We—this generation of humans—are at our most important juncture since we came out of the trees six million years ago. It is our decision, ours today, whether Earth continues to be a marvelously living, diverse oasis in the blackness of space, or whether the "charismatic megafauna" of the future will consist of Norway Rats and cockroaches.

How have we arrived at this state, at this threshold of biotic terror? Is it because we have forgotten our "place in nature," as the Native American activist Russell Means says?

If there is one thing upon which the nation states of the world today can agree, one thing at which the United States and the Soviet Union, Israel and Iran, South Africa and Angola, Britain and Argentina, China and India, Japan and Malaysia nod in unison, it is that human beings are the measure of all value. As Gifford Pinchot, founder of the United States Forest Service, said, there are only two things on Earth: human beings and natural resources. Humanism is the philosophy that runs the business engines of the modern world.

The picture that most humans have of the natural world is that of a smorgasbord table, continually replenished by a magic kitchen hidden somewhere in the background. While most people perceive that there are gross and immoral inequities in the sizes of the plates handed out and in the number of times some are allowed to belly up to the bar, few of us question whether the items arrayed are there for their sole use, nor do they imagine that the table will ever become empty.

There is another way to think about man's relationship to the natural world, an insight pioneered by the nineteenth-century conservationist and mountaineer John Muir and later by the science of ecology. This is the idea that all things are connected, interrelated, that human beings are merely one of the millions of species that have been shaped by the process of evolution for three and a half billion years. According to this view, all living beings have the same right to be here. This is how I see the world.

With that understanding, we can answer the question, "Why wilderness?"

Is it because wilderness makes pretty picture postcards? Because it protects watersheds for downstream use by agriculture, industry, and homes? Because it's a good place to clean the cobwebs out of our heads after a long week in the auto factory or over the video display terminal? Because it preserves resource-extraction opportunities for future generations of humans? Because some unknown plant living in the wilds may hold a cure for cancer?

No—the answer is, because wilderness *is*. Because it is the real world, the flow of life, the process of evolution, the repository of that three and a half billion years of shared travel.

A Grizzly Bear snuffling along Pelican Creek in Yellowstone National Park with her two cubs has just as much right to life as any human has, and is far more important ecologically. All things have intrinsic value, inherent worth. Their value is not determined by what they will ring up on the cash register of the gross national product, or by whether or not they are *good*. They are good because they exist.

Even more important than the individual wild creature is the wild community—the wilderness, the stream of life unimpeded by human manipulation.

We, as human beings, as members of industrial civilization, have no divine mandate to pave, conquer, control, develop, or use every square inch of this planet. As Edward Abbey, author of *Desert Solitaire* and *The Monkey Wrench Gang*, said, we have a right to be here, yes, but not everywhere, all at once.

The preservation of wilderness is not simply a question of balancing competing special-interest groups, arriving at a proper mix of uses on our public lands, and resolving conflicts between different outdoor recreation preferences. It is an ethical and moral matter. A religious mandate. Human beings have stepped beyond the bounds; we are destroying the very process of life.

The forest ranger and wilderness proponent Aldo Leopold perhaps stated this ethic best:

A thing is right when it tends to preserve the integrity, stability, and beauty of the biotic community. It is wrong when it tends otherwise. [1]

The crisis we now face calls for *passion*. When I worked as a conservation lobbyist in Washington, D.C., I was told to put my heart in a safe deposit box and replace my brain with a pocket calculator. I was told to be rational, not emotional, to use facts and figures, to quote economists and scientists. I would lose credibility, I was told, if I let my emotions show.

But, damn it, I am an animal. A living being of flesh and blood, storm and fury. The oceans of the Earth course through my veins, the winds of the sky fill my lungs, the very bedrock of the planet

---

1. All of the quotations from Aldo Leopold used in this book come from his classic work *A Sand County Almanac* (New York: Oxford University Press, 1949), which is widely considered to be the most important conservation book of this century. Leopold is probably the most influential conservation thinker our country has produced. For a fine biography of him, see *Aldo Leopold: His Life and Work*, by Curt Meine (Madison: University of Wisconsin Press, 1988).

makes my bones. I am alive! I am not a machine, a mindless automaton, a cog in the industrial world, some New Age android. When a chain saw slices into the heartwood of a two-thousand-year-old Coast Redwood, it's slicing into my guts. When a bulldozer rips through the Amazon rain forest, it's ripping into my side. When a Japanese whaler fires an exploding harpoon into a great whale, my heart is blown to smithereens. I am the land, the land is me.

Why shouldn't I be emotional, angry, passionate? Madmen and madwomen are wrecking this beautiful, blue-green, living Earth. Fiends who hold nothing of value but a greasy dollar bill are tearing down the pillars of evolution a-building for nearly four thousand million years.

In this world ruled by MBAs, we are taught to use only a fraction of our minds: the left hemisphere of the brain, the rational, calculating part. That portion of our brain is valuable and necessary, but it is not the sole seat of our consciousness. We must get back in touch with the emotional, intuitive right hemisphere of our brain, with our reptilian cortex, with our entire body. Then we must go beyond that to think with the whole Earth. David Brower, onetime executive director of the Sierra Club, has pointed out that you cannot imprison a California Condor in the San Diego Zoo and still have a condor. The being of a condor does not end at the tips of the black feathers on its wings. The condor is *place* as well; it is the thermals rising over the Coast Range, the outcroppings on which it lays its eggs, the carrion on which it feeds.

Society has lobotomized us. Our social environment today can work as a drug, like *soma* in *Brave New World,* to keep us in line, to sedate us, to remove our capacity for passion. Robots do not ask questions. Free men and women do. Wild animals cannot be ruled; they can be domesticated, yes, they can be broken, but then they are no longer free, no longer wild.

We must break out of society's freeze on our passions, we must become animals again. We must feel the tug of the full moon, hear goose music overhead. We must love Earth and rage against her destroyers. We must open ourselves to relationships with one another, with the land; we must dare to love, to feel for something—

some*one*—else. And when that final kiss of life—death—comes, we mustn't hide, but rather go *joyously* into that good night. When I die, I don't want to be pickled and put away in a lead box. Place me out in the wilderness, let me revel at rejoining the food chain, at being recycled into weasel, vulture, worm, and mold.

Breaking free from the gilded chains of civilized banality is not easy. One cannot achieve a state of wilderness grace through books, through intellectualization, through rational argument. Our passion comes from our connection to the Earth and it is only through direct interaction with the wilderness that we can unite our minds and our bodies with the land, realizing that there is no separation.

Along with passion, we need *vision*. Why should we content ourselves with the world the way it is handed to us by Louisiana-Pacific, Mitsubishi, the Pentagon, and Exxon? Why should we be constrained by the narrow alternatives presented us by Congress and the Forest Service in discussing protection of the land?

We are told that the Gray Wolf and Grizzly Bear are gone from most of the West and can never be restored, that the Elk and Bison and Panther are but shades in the East and will not come back, that Glen Canyon and Hetch Hetchy are beneath dead reservoir water and we shall never see them again, that the Tall Grass Prairie and Eastern Deciduous Forest* are only memories and that we can never have big wilderness east of the Rockies again.

---

*Throughout this book I have capitalized the proper names of species. Therefore, "Gray Wolf" or "Ponderosa Pine," being distinct species, are capitalized; "wolf" or "pine," being generic terms, are not. Similarly, I capitalize the formal name of a specific ecosystem like "Tall Grass Prairie" but not a generic term like "prairie." I capitalize "Wilderness" when it refers to formal, congressionally designated Wilderness Areas; I leave "wilderness" uncapitalized when it refers to de facto or undesignated wilderness. I capitalize "National Park," "National Wildlife Refuge," and "National Forest" whether they refer to a specific unit (e.g., Grand Canyon National Park) or simply to a generalized area or concept. Capitalization denotes a form of respect, and formal species' names, ecosystems, and preservation systems deserve that respect. It goes without saying that I therefore capitalize "Earth."

Bunk! Why should we be bound by past mistakes? It is up to us to challenge the government and the people with a vision of Big Wilderness, a vision of humans living modestly in a community that also includes bears and rattlesnakes and salmon and oaks and sagebrush and mosquitoes and algae and streams and rocks and clouds.

We should demand that roads be closed and clearcuts rehabilitated, that dams be torn down, that wolves, Grizzlies, Cougars, River Otters, Bison, Elk, Pronghorn, Bighorn Sheep, Caribou, and other extirpated species be reintroduced to their native habitats. We must envision and propose the restoration of biological wildernesses of several million acres in all of America's ecosystems, with corridors between them for the transmission of genetic variability. Wilderness is the arena for evolution, and there must be enough of it for natural forces to have free rein.

John Seed, the Australian founder of the Rainforest Information Centre, tells of a meeting he had with a group of Australian Aborigines in Sydney. After the meeting, they stepped outside into the night air. The great city spread out before them. One of the Abos asked, "What do you see? What do you see out *there*?"

John looked at the pulsating freeways, towers of anodized glass and steel, ships in the harbor, and replied, "I see a city. Lights, pavement, skyscrapers . . ."

The Abo said quietly, "We still see the land. Beneath the concrete we know where the forest grows, where the kangaroos graze. We see where the Platypus digs her den, where the streams flow. That city there . . . it's just a scab. The land remains alive beneath it."

So it is in North America. In the scrub forests of New England, the spirits of 220-foot-tall White Pines still stand. In the feedlots and cornfields of the Great Plains, ghost hooves of Bison and howls of wolves echo back from a century ago. On San Francisco shores, phantom Grizzlies feed on the beached carcasses of whales.

The genocide against these wilderness nations waged around the world by civilized humans has been going on for only an instant in evolutionary time. Some species are gone forever, some ecosystems are hopelessly muddled, but in most cases the land, the wild land, is still alive beneath the scab of concrete. Do we have the sight to see?

Passion and vision are essential, but without *action* they are empty. It is easy to be immobilized by the sheer magnitude of the problems facing Earth, by tasks calling for Hercules when we know we are puny mortals. We feel daunted about demanding changes when we know that our lives are not pure, that we share the life-style that is ravaging the planet. We feel powerless in confronting the vast, immobile gray bureaucracy of government and industry.

"It's too much," we whimper, and surrender. "Better not to fight than to be defeated. Besides, where does one person start? I'm not an expert or a leader. Why don't *they* do something?"

We are frozen because the problems are too big. It's easier to turn on the TV, to plunge into the modern game (whoever dies with the most toys, wins!), to dull our expectations and our passions with drink or with lines of white powder.

The Earth is crying. Do we hear? Martin Luther King, Jr., once said that if one has nothing worth dying for, one has nothing worth living for.

It is a time for courage.

There are many forms of courage. It takes courage to not allow your children to become addicted to television. It takes courage to tell the conservation group to which you belong, *No more compromise!* It takes courage to say no more growth in your community. It takes courage to say that the wild is more important than jobs. It takes courage to write letters to your local newspaper. It takes courage to stand up at a public hearing and speak. It takes courage to live a lower-impact life.

And it takes courage to put your body between the machine and the wilderness, to stand before the chain saw or the bulldozer.

In 1848, Henry David Thoreau went to jail for refusing, as a protest against the Mexican War, to pay his poll tax. When Ralph Waldo Emerson came to bail him out, Emerson said, "Henry, what are you doing in there?"

Thoreau quietly replied, "Ralph, what are you doing out there?"

In this insane world where short-term greed rules over long-term life, those of us with a land ethic, with vision and passion, must face the mad machine. We must stand before it as 19-year-old Oregon Earth First! activist Valerie Wade did when she climbed eighty feet up into an ancient Douglas-fir to keep it from being cut down; as Wyoming guide and outfitter and Earth First! founder Howie Wolke did when he pulled up survey stakes along a proposed gas-exploration road in prime Elk habitat. Both put their lives in jeopardy, both went to jail. Both were proud of what they did. Both are heroes of the Earth, as are hundreds of others who have demonstrated courage in defense of the wild.

This defense is not an arrogant defense, an attitude of Lord Man protecting something less than himself. Rather, it is a humble joining with Earth, becoming the rain forest, the desert, the mountain, the wilderness in defense of yourself. It is through becoming part of the wild that we find courage far greater than ourselves, a union that gives us boldness to stand against hostile humanism, against the machine, against the dollar, against jail, against extinction for what is sacred and right: the Great Dance of Life.

Eighty years ago, Aldo Leopold graduated from the Yale School of Forestry and went to work for the newly created United States Forest Service in the territories of Arizona and New Mexico. He was put to work inventorying potential timber resources in the high, wild White Mountains of eastern Arizona, which were a great roadless area then. One day Leopold stopped for lunch with his crew on a rimrock overlooking a turbulent stream. As they ate, they saw a large animal ford the *rillito*. They thought at first it was a doe, but as a rolling bunch of pups came out of the willows to greet their mother, they realized it was a wolf. In those days, a wolf you saw was a wolf you shot. Leopold and his men hurriedly pulled their .30-30s from the scabbards on their horses and began to blast away. The wolf dropped, a pup dragged a shattered leg into the rocks, and Leopold rode down to finish the job. He later wrote:

We reached the old wolf in time to watch a fierce green fire dying in her eyes. I realized then, and have known ever since, that there was something new to me in those eyes—something known only to her and to the mountain. I was young then, and full of trigger itch; I thought that because fewer wolves meant more deer, that no wolves would mean hunters' paradise. But after seeing the green fire die, I sensed that neither the wolf nor the mountain agreed with such a view.

Green fire. We need it in the eyes of the wolf. We need it in the land. And we need it in our own eyes.

2 | Earth First!

If opposition is not enough, we must resist. And if resistance is not enough, then subvert.

—*Edward Abbey*

The early conservation movement in the United States was a child—and no bastard child—of the Establishment. The founders of the Sierra Club, the National Audubon Society, The Wilderness Society, and the wildlife conservation groups were, as a rule, pillars of American society. They were an elite band—sportsmen of the Teddy Roosevelt variety, naturalists like John Burroughs, outdoorsmen in the mold of John Muir, pioneer foresters and ecologists on the order of Aldo Leopold, and wealthy social reformers like Gifford Pinchot and Robert Marshall. No anarchistic Luddites, these.

When the Sierra Club grew into the politically effective force that blocked Echo Park Dam in 1956 and got the Wilderness Act passed in 1964, its members (and members of like-minded organizations) were likely to be physicians, mathematicians, and nuclear physicists. To be sure, refugees from the mainstream joined the conservation outfits in the 1950s and 1960s, and David Brower, executive director of the Sierra Club during that period, and the man most responsible for the creation of the modern environmental movement, was begin-

11

ning to ask serious questions about the assumptions and direction of industrial society by the time the Club's board of directors fired him in 1969. But it was not until Earth Day in 1970 that the environmental movement received its first real influx of antiestablishment radicals, as Vietnam War protestors found a new cause—the environment. Suddenly, beards appeared alongside crewcuts in conservation group meetings—and the rhetoric quickened.

The militancy was short-lived. Eco-anarchist groups like Black Mesa Defense, which provided a cutting edge for the movement, peaked at the United Nations' 1972 Stockholm Conference on the Human Environment, but then faded from the scene. Along with dozens of other products of the 1960s who went to work for conservation organizations in the early 1970s, I discovered that compromise seemed to work best. A suit and a tie gained access to regional heads of the U.S. Forest Service and to members of Congress. We learned to moderate our opinions along with our dress. We learned that extremists were ignored in the councils of government, that the way to get a senator to put his arm around your shoulders and drop a Wilderness bill in the hopper was to consider the conflicts—mining, timber, grazing—and pare back the proposal accordingly. *Of course* we were good, patriotic Americans. *Of course* we were concerned with the production of red meat, timber, and minerals. We tried to demonstrate that preserving wilderness did not conflict all that much with the gross national product, and that clean air actually helped the economy. We argued that we could have our booming industry and still not sink oil wells in pristine areas.

This moderate stance appeared to pay off when Jimmy Carter, the first President who was an avowed conservationist since Teddy Roosevelt, took the helm at the White House in 1977. Self-professed conservationists were given decisive positions in Carter's administration. Editorials proclaimed that environmentalism had been enshrined in the Establishment, that conservation was here to stay. A new ethic was at hand: Environmental Quality and Continued Economic Progress.

Yet, although we had access to and influence in high places, something seemed amiss. When the chips were down, conservation still lost out to industry. But these were our friends turning us down. We tried to understand the problems they faced in the real political world. We gave them the benefit of the doubt. We failed to sue when we should have. . . .

I wondered about all this on a gray day in January 1979 as I sat in my office in the headquarters of The Wilderness Society, only three blocks from the White House. I had just returned from a news conference at the South Agriculture Building, where the Forest Service had announced a disappointing decision on the second Roadless Area Review and Evaluation, a twenty-month exercise by the Forest Service to determine which National Forest lands should be protected in their natural condition. As I loosened my tie, propped my cowboy boots up on my desk, and popped the top on another Stroh's, I thought about RARE II and why it had gone so wrong. Jimmy Carter was supposedly a great friend of wilderness. Dr. M. Rupert Cutler, a former assistant executive director of The Wilderness Society, was Assistant Secretary of Agriculture over the Forest Service and had conceived the RARE II program. But we had lost to the timber, mining, and cattle interests on every point. Of 80 million acres still roadless and undeveloped in the 190 million acres of National Forests, the Department of Agriculture recommended that only 15 million receive protection from road building and timber cutting.[1] Moreover, damn it, we—the conservationists—had been moderate. The antienvironmental side had been extreme, radical, emotional, their arguments full of holes. We had been factual, rational. We had provided more—and better—serious public comment. But we had lost, and now we were worried that some local wilderness group might go off the reservation and sue the Forest Service over the clearly inadequate environmental impact statement for RARE II. We didn't want a lawsuit because we knew we could win and were afraid of the political consequences of such

1. Only 62 million acres were actually considered by the Forest Service in RARE II. Another 18 million acres that were also roadless and undeveloped were not considered because of sloppy inventory procedures, political pressure, or because areas had already gone through land-use plans that had supposedly considered their wilderness potential.

a victory. We might make some powerful senators and representatives angry. So those of us in Washington were plotting how to keep the grassroots in line. Something about all this seemed wrong to me.

After RARE II, I left my position as issues coordinator for The Wilderness Society in Washington to return to New Mexico and my old job as the Society's Southwest representative. I was particularly concerned with overgrazing on the 180 million acres of public lands in the West managed by the Department of the Interior's Bureau of Land Management. For years, these lands—rich in wildlife, scenery, recreation, and wilderness—had been the private preserve of stock growers in the West. BLM had done little to manage the public lands or to control the serious overgrazing that was sending millions of tons of topsoil down the Colorado, the Rio Grande, and other rivers, wiping out wildlife habitat, and generally beating the land to hell.

Prodded by a Natural Resources Defense Council lawsuit, BLM began to address the overgrazing problem through a series of environmental impact statements. These confirmed that most BLM lands were seriously overgrazed, and recommended cuts in livestock numbers. But after the expected outcry from ranchers and their political cronies in Congress and in state capitals, BLM backtracked so quickly that the Department of the Interior building suffered structural damage. Why were BLM and the Department of the Interior so gutless?

While that question gnawed at my innards, I was growing increasingly disturbed about trends in the conservation organizations themselves. When I originally went to work for The Wilderness Society in 1973, the way to get a job with a conservation group was to prove yourself first as a volunteer. It helped to have the right academic background, but experience as a capable grassroots conservation activist was more important.

We realized that we would not receive the salaries we could earn in government or private industry, but we didn't expect them. We were working for nonprofit groups funded by the contributions of concerned people. Give us enough to keep food on the table, pay rent, buy a six-pack—we didn't want to get rich. But a change occurred after the mid-1970s: people seeking to work for conservation groups were career-oriented; they had relevant degrees (science or law, not history or English); they saw jobs in environmental organizations in the same light as jobs in government or industry. One was a steppingstone to another, more powerful position later on. They were less part of a cause than members of a profession.

A gulf began to grow between staff and volunteers. We also began to squabble over salaries. We were no longer content with subsistence wages, and the figures on our paychecks came to mark our status in the movement. Perrier and Brie replaced Bud and beans at gatherings.

Within The Wilderness Society, executive director Celia Hunter, a prominent Alaskan conservationist and outfitter, World War II pilot, and feminist, was replaced in 1978 by Bill Turnage, an eager young businessman who had made his mark by marketing Ansel Adams. Within two years Turnage had replaced virtually all those on the staff under Celia with professional organization people. The governing council also worked to bring millionaires with a vague environmental interest on board. We were, it seemed to some of us, becoming indistinguishable from those we were ostensibly fighting.

I resigned my position in June 1980.

The same dynamics seemed to affect the rest of the movement. Were there any radicals anywhere? Anyone to take the hard stands? Sadly, no. The national groups—Sierra Club, Friends of the Earth, National Audubon Society, Natural Resources Defense Council, and the rest—took almost identical middle-of-the-road positions on most issues. And then those half-a-loaf demands were readily compromised further. The staffs of these groups fretted about keeping local conservationists (and some of their field representatives) in line, keeping them from becoming extreme or unreasonable, keeping them from blowing moderate national strategy. Even Friends of the Earth, which had started out radical back in the heady Earth Day era, had gravitated to the center and, as a rule, was a comfortable member of the informal coalition of big environmental organizations.

For years I advocated this approach. We could, I believed, gain more wilderness by taking a moderate tack. We would stir up less opposition by keeping a low profile. We could inculcate conservation in the Establishment by using rational economic arguments. We needed to present a solid front.

A major crack in my moderate ideas appeared early in 1979, when I returned from Washington to the small ranching community of Glenwood, New Mexico. I had lived there earlier for six years, and although I was a known conservationist, I was fairly well accepted. Shortly after my return, *The New York Times* published an article on RARE II, with the Gila National Forest around Glenwood as chief exhibit. To my amazement, the article quoted a rancher who I considered to be a friend as threatening *my* life because of local fears about the consequences of wilderness designations. A couple of days later I was accosted on the street by four men, one of whom ran the town café where I had eaten many a chicken-fried steak. They threatened my life because of RARE II.

I was not afraid, but I was irritated—and surprised. I had been a leading moderate among New Mexico conservationists. I had successfully persuaded New Mexico conservation groups to propose fewer RARE II areas on the Gila National Forest as Wilderness. What had backfired? I thought again about the different approaches to RARE II: the moderate, subdued one advanced by the major conservation groups; the howling, impassioned, extreme stand set forth by off-road-vehicle zealots, many ranchers, local boosters, loggers, and miners. They looked like fools. We looked like statesmen. They won.

The last straw fell on the Fourth of July, 1980, in Moab, Utah. There the local county commission sent a flag-flying bulldozer into an area the Bureau of Land Management had identified as a possible study area for Wilderness designation. The bulldozer incursion was an opening salvo for the so-called Sagebrush Rebellion, a move by chambers of commerce, ranchers, and right-wing fanatics in the West to claim federal public land for the states and eventual transfer to private hands. The Rebellion was clearly an extremist effort, lacking the support of even many conservative members of Congress in the West, yet BLM was afraid to stop the county commission.

What have we really accomplished? I thought. *Are we any better off as far as saving the Earth now than we were ten years ago?* I ticked off the real problems: world population growth, destruction of tropical forests, expanding slaughter of African wildlife, oil pollution of the oceans, acid rain, carbon dioxide buildup in the atmosphere, spreading deserts on every continent, destruction of native peoples and the imposition of a single culture (European) on the entire world, plans to carve up Antarctica, planned deep seabed mining, nuclear proliferation, recombinant DNA research, toxic wastes. . . . It was staggering. And I feared we had done nothing to reverse the tide. Indeed, it had accelerated.

And then: Ronald Reagan. James "Rape 'n' Ruin" Watt became Secretary of the Interior. The Forest Service was Louisiana-Pacific's. Interior was Exxon's. The Environmental Protection Agency was Dow's. Quickly, the Reagan administration and the Republican Senate spoke of gutting the already gutless Alaska Lands bill. The Clean Air Act, up for renewal, faced a government more interested in corporate black ink than human black lungs. The lands of the Bureau of Land Management appeared to the Interior Department obscenely naked without the garb of oil wells. Concurrently, the Agriculture Department directed the Forest Service to rid the National Forests of decadent and diseased old-growth trees. The cowboys had the grazing lands, and God help the hiker, Coyote, or blade of grass that got in their way.

Maybe, some of us began to feel, even before Reagan's election, it was time for a new joker in the deck: a militant, uncompromising group unafraid to say what needed to be said or to back it up with stronger actions than the established organizations were willing to take. This idea had been kicking around for a couple of years. Finally, in 1980, several disgruntled conservationists—including Susan Morgan, formerly educational director for The Wilderness Society; Howie Wolke, former Wyoming representative for Friends of the Earth; Bart Koehler, former Wyoming representative for The Wilderness Society; Ron Kezar, a longtime Sierra Club activist; and I—decided that the time for talk was past. We formed a new national group, which we called Earth First! We set out to be radical in style, positions, philosophy, and organization in order to be effective and to avoid the pitfalls of co-option and moderation that we had already experienced.

What, we asked ourselves as we sat around a campfire in the Wyoming mountains, were the reasons and purposes for environmental radicalism?

¶To state honestly the views held by many conservationists.

¶To demonstrate that the Sierra Club and its allies were raging moderates, believers in the system, and to refute the Reagan/Watt contention that they were "environmental extremists."

¶To balance such antienvironmental radicals as the Grand County commission and provide a broader spectrum of viewpoints.

¶To return vigor, joy, and enthusiasm to the tired, unimaginative environmental movement.

¶To keep the established groups honest. By stating a pure, no-compromise, pro-Earth position, we felt that Earth First! could help keep the other groups from straying too far from their original philosophical base.

¶To give an outlet to many hard-line conservationists who were no longer active because of disenchantment with compromise politics and the co-option of environmental organizations.

¶To provide a productive fringe, since ideas, creativity, and energy tend to spring up on the edge and later spread into the center.

¶To inspire others to carry out activities straight from the pages of *The Monkey Wrench Gang* (a novel of environmental sabotage by Edward Abbey), even though Earth First!, we agreed, would itself be ostensibly law-abiding.

¶To help develop a new worldview, a biocentric paradigm, an Earth philosophy. To fight, with uncompromising passion, for Earth.

The name Earth First! was chosen because it succinctly summed up the one thing on which we could all agree: That in *any* decision, consideration for the health of the Earth must come first.

In a true Earth-radical group, concern for wilderness preservation must be the keystone. The idea of wilderness, after all, is the most radical in human thought—more radical than Paine, than Marx, than Mao. Wilderness says: Human beings are not paramount, Earth is not for *Homo sapiens* alone, human life is but one life form on the planet and has no right to take exclusive possession. Yes, wilderness for its own sake, without any need to justify it for human benefit. Wilderness for wilderness. For bears and whales and titmice and rattlesnakes and stink bugs. And . . . wilderness for human beings. Because it is the laboratory of human evolution, and because it is home.

It is not enough to protect our few remaining bits of wilderness. The only hope for Earth (including humanity) is to withdraw huge areas as inviolate natural sanctuaries from the depredations of modern industry and technology. Keep Cleveland, Los Angeles. Contain them. Try to make them habitable. But identify big areas that can be restored to a semblance of natural conditions, reintroduce the Grizzly Bear and wolf and prairie grasses, and declare them off limits to modern civilization.

In the United States, pick an area for each of our major ecosystems and recreate the American wilderness—not in little pieces of a thousand acres, but in chunks of a million or ten million. Move out the people and cars. Reclaim the roads and plowed land. It is not enough any longer to say no more dams on our wild rivers. We must begin tearing down some dams already built—beginning with Glen Canyon on the Colorado River in Arizona, Tellico in Tennessee, Hetch Hetchy and New Melones in California—and freeing shackled rivers.

This emphasis on wilderness does not require ignoring other environmental issues or abandoning social issues. In the United States, blacks and Chicanos of the inner cities are the ones most affected by air and water pollution, the ones most trapped by the unnatural confines of urbanity. So we decided that not only should eco-militants be concerned with these human environmental problems, we should also make common ground with other progressive elements of society whenever possible.

Obviously, for a group more committed to Gila Monsters and Mountain Lions than to people, there will not be a total alliance with other social movements. But there are issues in which Earth radicals can cooperate with feminist, Native American, anti-nuke, peace, civil-rights, and civil-liberties groups. The inherent conservatism of the conservation community has made it wary of snuggling too close to these leftist organizations. We hoped to pave the way for better cooperation from the entire conservation movement.

We believed that new tactics were needed—something more than commenting on dreary environmental-impact statements and writing letters to members of Congress. Politics in the streets. Civil disobedience. Media stunts. Holding the villains up to ridicule. Using music to charge the cause.

Action is the key. Action is more important than philosophical hairsplitting or endless refining of dogma (for which radicals are so well known). Let our actions set the finer points of our philosophy. And let us recognize that diversity is not only the spice of life, but also the strength. All that would be required to join Earth First!, we decided, was a belief in Earth first. Apart from that, Earth First! would be big enough to contain street poets and cowboy bar bouncers, agnostics and pagans, vegetarians and raw-steak eaters, pacifists and those who think that turning the other cheek is a good way to get a sore face.

Radicals frequently verge on a righteous seriousness. But we felt that if we couldn't laugh at ourselves we would be merely another bunch of dangerous fanatics who should be locked up—like oil company executives. Not only does humor preserve individual and group sanity; it retards hubris, a major cause of environmental rape, and it is also an effective weapon. Fire, passion, courage, and emotionalism are also needed. We have been too reasonable, too calm, too understanding. It's time to get angry, to cry, to let rage flow at what the human cancer is doing to Earth, to be uncompromising.

For Earth First! there is no truce or cease-fire. No surrender. No partitioning of the territory.

Ever since the Earth goddesses of ancient Greece were supplanted by the macho Olympians, repression of women and Earth has gone hand in hand with imperial organization. Earth First! decided to be nonorganizational: no officers, no bylaws or constitution, no incorporation, no tax status, just a collection of women and men committed to the Earth. At the turn of the century, William Graham Sumner wrote a famous essay titled "The Conquest of the United States by Spain." His thesis was that Spain had ultimately won the Spanish-American War because the United States took on the imperialism and totalitarianism of Spain. We felt that if we took on the organization of the industrial state, we would soon accept their anthropocentric paradigm, much as Audubon and the Sierra Club already had.

And when we are inspired, we *act.*

Massive, powerful, like some creation of Darth Vader, Glen Canyon Dam squats in the canyon of the Colorado River on the Arizona-Utah border and backs the cold, dead waters of "Lake" Powell some 180 miles upstream, drowning the most awesome and magical canyon on Earth. More than any other single entity, Glen Canyon Dam is the symbol of the destruction of wilderness, of the technological ravishment of the West. The finest fantasy of eco-warriors in the West is the destruction of the dam and the liberation of the Colorado. So it was only proper that on March 21, 1981—at the spring equinox, the traditional time of rebirth—Earth First! held its first national gathering at Glen Canyon Dam.

On that morning, seventy-five members of Earth First! lined the walkway of the Colorado River Bridge, seven hundred feet above the once-free river, and watched five compatriots busy at work with an awkward black bundle on the massive dam just upstream. Those on the bridge carried placards reading "Damn Watt, Not Rivers," "Free the Colorado," and "Let it Flow." The five of us on the dam

DAVE FOREMAN

attached ropes to a grille, shouted out "Earth First!" and let three hundred feet of black plastic unfurl down the side of the dam, creating the impression of a growing crack. Those on the bridge returned the cheer.

Then Edward Abbey, author of *The Monkey Wrench Gang*, told the protestors of the "green and living wilderness" that was Glen Canyon only nineteen years ago:

> *And they took it away from us. The politicians of Arizona, Utah, New Mexico, and Colorado, in cahoots with the land developers, city developers, industrial developers of the Southwest, stole this treasure from us in order to pursue and promote their crackpot ideology of growth, profit, and power—growth for the sake of power, power for the sake of growth.*

Speaking toward the future, Abbey offered this advice: "Oppose. Oppose the destruction of our homeland by these alien forces from Houston, Tokyo, Manhattan, Washington, D.C., and the Pentagon. And if opposition is not enough, we must resist. And if resistance is not enough, then subvert."

Hardly had he finished speaking when Park Service police and Coconino County sheriff's deputies arrived on the scene. While they questioned Howie Wolke and me, and tried to disperse the illegal assembly, outlaw country singer Johnny Sagebrush led the demonstrators in song for another twenty minutes.

The Glen Canyon Dam caper brought Earth First! an unexpected amount of media attention. Membership quickly spiraled to more than a thousand, with members from Maine to Hawaii. Even the government became interested. According to reports from friendly park rangers, the FBI dusted the entire Glen Canyon Dam crack for fingerprints!

When a few of us kicked off Earth First!, we sensed a growing environmental radicalism in the country, but we did not expect the response we received. Maybe Earth First! is in the right place at the right time.

The cynical may smirk. "But what can you really accomplish? How can you fight Exxon, Coors, the World Bank, Japan, and the

22

other great corporate giants of the Earth? How, indeed, can you fight the dominant dogmas of Western civilization?"

Perhaps it *is* a hopeless quest. But one who loves Earth can do no less. Maybe a species will be saved or a forest will go uncut or a dam will be torn down. Maybe not. A monkeywrench thrown into the gears of the machine may not stop it. But it might delay it, make it cost more. And it feels good to put it there.

3 | Putting the Earth First

These are the times that try men's souls; the summer soldier and the sunshine patriot will, in this crisis, shrink from the service of his country, but he that stands it now, deserves the love and thanks of man and woman.

—Thomas Paine

In July 1987, seven years after the campfire gathering that spawned Earth First!, I rose among the Ponderosa Pines and scattered shafts of sunlight on the North Rim of the Grand Canyon and mounted a stage festooned with Earth First! banners and American flags. Before me sat several hundred people: hippies in tie-dyed shirts and Birkenstocks, rednecks for wilderness in cowboy boots and hats, middle-class hikers in waffle stompers, graybeards and children. The diversity was impressive. The energy was over-powering. Never in my wildest dreams had I imagined the Earth First! movement would attract so many. Never had I hoped that we would have begun to pack such a punch. We were attracting national attention; we were changing the parameters of the debate about ecological issues; we had become a legend in conservation lore.

Yet, after seven years, I was concerned we were losing some of our clarity of purpose, and blurring our focus. In launching Earth First!, I had said, "Let our actions set the finer points of our philosophy." But now I was concerned that the *what* of our actions might be overwhelming the *why*. For some of those newly attracted to

Earth First!, action seemed to be its own justification. I felt a need to return to wilderness fundamentalism, to articulate what I thought were the principles that defined the Earth First! movement, that gave it a specific identity. The response to the principles I offered that day was so overwhelmingly positive that I elaborated on them in the *Earth First! Journal* later that fall. Here they are.

A placing of Earth first in all decisions, even ahead of human welfare if necessary. Our movement is called "Earth First!" not "People First!" Sometimes what appears to be in the short-term interest of human beings as a whole, a select group of human beings, or individual human beings is detrimental to the short-term or long-term health of the biosphere (and to the actual long-term welfare of human beings). Earth First! does not argue that native diversity should be preserved if it can be done without negatively impacting the material "standard of living" of a group of human beings. We simply state that native diversity should be preserved, that natural diversity a-building for three and a half billion years should be left unfettered. Human beings must adjust to the planet; it is supreme arrogance to expect the planet and all it contains to adjust to the demands of humans. In everything human society does, the primary consideration should be for the long-term health and biological diversity of Earth. After that, we can consider the welfare of humans. We should be kind, compassionate, and caring with other people, but Earth comes first.

A refusal to use human beings as the measure by which to value others. An individual human life has no more intrinsic value than does an individual Grizzly Bear life. Human suffering resulting from drought and famine in Ethiopia is tragic, yes, but the destruction there of other creatures and habitat is even more tragic. This leads quickly into the next point:

An enthusiastic embracing of the philosophy of Deep Ecology or biocentrism. This philosophy states simply and essentially that all living creatures and communities possess intrinsic value, inherent worth. Natural things live for their own sake, which is another way of saying they have value. Other beings (both animal and plant) and even so-called "inanimate" objects such as rivers and mountains are not placed here for the convenience of human beings. Our biocentric worldview denies the modern concept of "resources." The dominant philosophy of our time (which contains Judeo-Christianity, Islam, capitalism, Marxism, scientism, and secular humanism) is anthropocentrism. It places human beings at the center of the universe, separates them from nature, and endows them with unique value. EF!ers are in direct opposition to that philosophy. Ours is an ecological perspective that views Earth as a community and recognizes such apparent enemies as "disease" (e.g., malaria) and "pests" (e.g., mosquitoes) not as manifestations of evil to be overcome but rather as vital and necessary components of a complex and vibrant biosphere.

A realization that wilderness is the real world. The preservation of wilderness is the fundamental issue. Wilderness does not merely mean backpacking parks or scenery. It is the natural world, the arena for evolution, the caldron from which humans emerged, the home of the others with whom we share this planet. Wilderness is the real world; our cities, our computers, our airplanes, our global business civilization all are but artificial and transient phenomena. It is important to remember that only a tiny portion of the history of the human species has occurred outside of wilderness. The preservation of wildness and native diversity is *the* most important issue. Issues directly affecting only humans pale in comparison. Of course, ecology teaches us that all things are connected, and in this regard all other matters become subsets of wilderness preservation—the prevention of nuclear war, for example—but the most important campaigns being waged today are those directly on behalf of wilderness.

A recognition that there are far too many human beings on Earth. There are too many of us everywhere—in the United States, in Nigeria; in cities, in rural areas; with digging hoes, with tractors. Although there is obviously an unconscionable maldistribution of wealth and the basic necessities of life among humans, this fact should not be used—as some leftists are wont to do—to argue that overpopulation is not the problem. It *is* a large part of the problem; there are far too many of us *already*—and our numbers continue to grow astronomically. Even if inequitable distribution could be solved, six billion human beings converting the natural world to material goods and human food would devastate natural diversity.

This basic recognition of the overpopulation problem does not mean that we should ignore the economic and social causes of overpopulation, and shouldn't criticize the accumulation of wealth in fewer and fewer hands, the maldistribution of "resources," and the venality of multinational corporations and Third World juntas alike, but simply that we must understand that Great Blue Whales, Jaguars, Black Rhinoceroses, and rain forests are not compatible with an exploding human population.[1]

A deep questioning of, and even an antipathy to, "progress" and "technology." In looking at human history, we can see that we have lost more in our "rise" to civilization than we have gained. We can see that life in a hunter-gatherer society was on the whole healthier, happier, and more secure than our lives today as peasants, industrial workers, or business executives. For every material "achievement" of progress, there are a dozen losses of things of profound and ineffable value. We can accept the pejoratives of "Luddite" and "Neanderthal" with pride. (This does not mean that we must immediately eschew all the facets of technological civilization. We are *of* it, and use it; this does not mean that we can't critique it.)

1. Two excellent books on the population issue that are also sensitive to social and economic issues are William R. Catton, Jr.'s *Overshoot: The Ecological Basis of Revolutionary Change* (Urbana, Ill., and Chicago: University of Illinois Press, 1982), and *The Population Explosion*, by Paul and Anne Ehrlich (New York: Simon and Schuster, 1990). No one concerned with the preservation of biological diversity should be without these.

A refusal to accept rationality as the only way of thinking. There is room for great diversity within Earth First! on matters spiritual, and nowhere is tolerance for diversity more necessary. But we can all recognize that linear, rational, logical left brain thinking represents only part of our brain and consciousness. Rationality is a fine and useful tool, but it is just that—a tool, one way of analyzing matters. Equally valid, perhaps more so, is intuitive, instinctive awareness. We can become more cognizant of ultimate truths by sitting quietly in the wild than by studying in a library. Reading books, engaging in logical discourse, and compiling facts and figures are necessary in the modern context, but they are not the only ways to comprehend the world and our lives. Often our gut instincts enable us to act more effectively in a crisis than does careful rational analysis. An example would be a patient bleeding to death in a hospital emergency room—you can't wait for all the tests to be completed. Your gut says, "Act!" So it is with Earth First!'s actions in Earth's current emergency.

A lack of desire to gain credibility or "legitimacy" with the gang of thugs running human civilization. It is basic human nature to want to be accepted by the social milieu in which you find yourself. It hurts to be dismissed by the arbiters of opinion as "nuts," "terrorists," "wackos," or "extremists." But we are not crazy; we happen to be sane humans in an insane human society in a sane natural world. We do not have "credibility" with Senator Mark Hatfield or with Maxxam chairman Charles Hurwitz—but they do not have credibility with us! (We do have their attention, however.) They are madmen destroying the pure and beautiful. Why should we "reason" with them? We do not share the same worldview or values. There is, however, a dangerous pitfall here that some alternative groups fall into. That is that we gain little by being consciously offensive, by trying to alienate others. We can be strong and unyielding without being obnoxious.

The American system is very effective at co-opting and moderating dissidents by giving them attention and then encouraging them to be "reasonable" so their ideas will be taken seriously. Putting a critic on the evening news, on the front page of the newspaper, in a national magazine—all of these are methods the establishment uses to entice one to share their worldview and to enter the negotiating room to compromise. The actions of Earth First!—both the bold and the comic—have gained attention. If they are to have results, we must resist the siren's offer of credibility, legitimacy, and a share in the decision-making. We are thwarting the system, not reforming it. While we are therefore not concerned with political credibility, it must be remembered that the arguments and actions of Earth First! are based on the understandings of ecology. It is vitally important that we have biological credibility.

An effort to go beyond the tired, worn-out dogmas of left, right, and middle-of-the-road. These doctrines, whether blaming capitalism, communism, or the devil for all the problems in the world, merely represent internecine squabbles between different factions of humanism. Yes, multinational corporations commit great evil (the Soviet Union is essentially a state-run multinational corporation); there is a great injustice in the world; the rich are getting richer and the poor poorer—but all problems cannot be simplistically laid at the feet of evil capitalists in the United States, Europe, and Japan. Earth First! is not left or right; we are not even in front. Earth First! should not be in the political struggle between humanist sects at all. We're in a wholly different game.

An unwillingness to set any ethnic, class, or political group of humans on a pedestal and make them immune from questioning. It's easy, of course, to recognize that white males from North America and Europe (as well as Japanese males) hold a disproportionate share of responsibility for the mess we're in; that upper- and middle-class consumers from the First World take an excessive portion of the world's "resources" and therefore cause greater per capita destruction than do other peoples. But it does not follow that everyone else is blameless.

The Earth First! movement has great affinity with aboriginal groups throughout the world. They are clearly in the most direct and respectful relationship with the natural world. Earth First! should back such tribes in the common struggle whenever possible without compromising our ideals. For example, we are supportive of the Dine (Navajo) of Big Mountain against relocation, but this does not mean we overlook the severe overgrazing by domestic sheep on the Navajo Reservation. We may be supportive of subsistence life-styles by natives in Alaska, but we should not be silent about clearcutting old-growth forest in southeast Alaska by native corporations, or about the Eskimo Doyon Corporation's push for oil exploration and development in the Arctic National Wildlife Refuge. It is racist either to condemn or to pardon someone based on their ethnic background.

Similarly, we are inconsistent when we castigate Charles Hurwitz for destroying the last wilderness redwood forest, yet feel sympathy for the loggers working for him. Industrial workers, by and large, share the blame for the destruction of the natural world. They may be yoked by the big-money boys, but they are generally willing servants who share the worldview of their bosses that Earth is a smorgasbord of resources for the taking. Sometimes, in fact, it is the sturdy yeoman from the bumpkin proletariat who holds the most violent and destructive attitudes toward the natural world (and toward those who would defend it).[2] Workers are victims of an unjust economic system, but that does not absolve them of what they do. This is not to deny that some woods workers oppose the destruction of ancient forests, that some may even be Earth First!ers, but merely that it is inappropriate to overlook abuse of the natural world simply because of the rung the perpetrators occupy on the economic ladder.

Some argue that workers are merely struggling to feed their families and are not delighting in destroying the natural world. They say that unless you deal with the needs of loggers to make a

2. A case in point involves the Spotted Owl, a Threatened species dependent on ancient forests. These little owls are easily attracted by playing tapes of their call. Loggers in the Northwest are going into old-growth forests with tape recorders and shotguns to exterminate Spotted Owls. They feel that if they do so, they will eliminate a major reason to stop the logging of these pristine forests.

living, you can't save the forest. They also claim that loggers are manipulated by their bosses to express anti-wilderness viewpoints. I find this argument to be patronizing to loggers and other workers. When I read comments from timber fellers expressing hatred toward pristine forests and toward conservationists, it is obvious that they willingly buy into the worldview of the lumber barons. San Francisco's *Image Magazine* reports on a letter to the editor written by one logger: "Working people trying to feed their families have little time to be out in the woods acting like children and making things hard for other working people. . . . Anyone out there have a recipe for spotted owl? Food stamps won't go far, I'm afraid. And since they're always being shoved down my throat, I thought I'd like mine fried."[3] Bumper stickers proclaiming "Kill an owl. Save a logger." are rife in the Northwest. I at least respect the logger who glories in felling a giant tree and who hunts Spotted Owls enough to grant him the mental ability to have his own opinions instead of pretending he is a stupid oaf, manipulated by his bosses and unable to think for himself.

Of course the big timber companies do manipulate their workers with scare tactics about mill closings and wilderness lockups, but many loggers (or cat-skinners, oilfield workers, miners, and the like) simply hate the wild and delight in "civilizing" it. Even educating workers about ecological principles will not necessarily change the attitudes of many; there are basic differences of opinion and values. Conservationists should try to find common ground with loggers and other workers whenever possible, but the sooner we get rid of Marxist views about the noble proletariat, the better.

A willingness to let our actions set the finer points of our philosophy and a recognition that we must act. It is possible to debate endlessly the finer points of dogma, to feel that every nuance of something must be explored before one can act. Too often, political movements become mere debating societies where

3. Jane Kay, "Tree Wars," *San Francisco Examiner Image Magazine* (December 17, 1989).

the participants engage in philosophical masturbation and never get down to the vital business at hand. Others argue that you have no right to argue for environmental preservation until you are living a pure, non-impacting life-style. We will never figure it all out, we will never be able to plan any campaign in complete detail, none of us will ever entirely transcend a polluting life-style—but we can act. We can act with courage, with determination, with love for things wild and free. We can't be perfect, but we can *act*. We are warriors. Earth First! is a warrior society. We have a job to do.

An acknowledgment that we must change our personal life-styles to make them more harmonious with natural diversity. We must eschew surplusage. Although to varying degrees we are all captives of our economic system and cannot break entirely free, we must practice what we preach to the best of our ability. Arne Naess, the Norwegian philosopher and originator of the term "Deep Ecology," points out that we are not able to achieve a true "Deep Ecology" life-style, but it is the responsibility of each of us to move in that direction. Most of us still need to make a living that involves some level of participation in "the system." Even for activists, there are trade-offs—flying in a jetliner to help hang a banner on the World Bank in Washington, D.C., in order to bring international attention to the plight of tropical rain forests; using a computer to write a book printed on tree pulp that will catalyze people to take action; driving a pickup truck down a forest road to gain access to a proposed timber sale for preventive maintenance. We need to be aware of these trade-offs, and to do our utmost to limit our impact.

A commitment to maintaining a sense of humor, and a joy in living. Most radicals are a dour, holier-than-thou, humorless lot. Earth First!ers strive to be different. We aren't rebelling against the system because we're losing in it. We're fighting for beauty, for life, for joy. We kick up our heels in delight in the wilderness, we smile at a flower and a hummingbird. We laugh. We laugh at our opponents—and, more important, we laugh at ourselves.

An awareness that we are animals. Human beings are primates, mammals, vertebrates. EF!ers recognize their animalness; we reject the New Age eco-la-la that says we must transcend our base animal nature and take charge of our evolution in order to become higher, moral beings. We believe we must return to being animal, to glorying in our sweat, hormones, tears, and blood. We struggle against the modern compulsion to become dull, passionless androids. We do not live sanitary, logical lives; we smell, taste, see, hear, and feel Earth; we live with gusto. We *are* Animal.

An acceptance of monkeywrenching as a legitimate tool for the preservation of natural diversity. Not all Earth First!ers monkeywrench, perhaps not even the majority, but we generally accept the idea and practice of monkeywrenching. Look at an EF! T-shirt. The monkeywrench on it is a symbol of resistance, an heir of the *sabot*—the wooden shoe dropped in the gears to stop the machine, from whence comes the word *sabotage*. The mystique and lore of "night work" pervades our tribe, and with it a general acceptance that strategic monkeywrenching is a legitimate tool for defense of the wild.

And finally: Earth First! is a warrior society. In addition to our absolute commitment to and love for this living planet, we are characterized by our willingness to defend Earth's abundance and diversity of life, even if that defense requires sacrifices of comfort, freedom, safety, or, ultimately, our lives. A warrior recognizes that her life is not the most important thing in her life. A warrior recognizes that there is a greater reality outside her life that must be defended. For us in Earth First!, that reality is Earth, the evolutionary process, the millions of other species with which we share this bright sphere in the void of space.

Not everyone can afford to make the commitment of being a warrior. There are many other roles that can—and must—be played in defense of Earth. One may not constantly be able to carry the burden of being a warrior; it may be only a brief period in one's life. There are risks and pitfalls in being a warrior. There may not be applause, there may not be honors and awards from human society. But there is no finer applause for the warrior of the Earth than the call of the loon at dusk or the sigh of wind in the pines.

Later that evening as I looked out over the darkening Grand Canyon, I knew that whatever hardships the future might bring, there was nothing better and more important for me to do than to take an intransigent stand in defense of life, to not compromise, to continue to be a warrior for the Earth. To be a warrior for the Earth regardless of the consequences.

4 | Who Speaks for Wolf?

All good things are wild, and free.

—Henry David Thoreau

O f all the myths generously marbled through the meat of American history, perhaps none is so cherished as that of "new beginnings."

I remember, in my earliest elementary-school history lessons, being taught that the colonists who left the Old World for the New—such as the Pilgrims and Puritans of New England, and the Quakers of Pennsylvania, and the later immigrants from eastern and southern Europe—came to find a new life, that they sought the freedom to develop a new society without the musty political baggage of Europe.

I come from a long line of these pioneers, of these *Americans.* My bones jelled out of the hillbilly bedrock of this nation. Although I number among my kin a signer of the Constitution and the wife of President John Tyler, my lineage is of common folk. Yes, my great-aunts in Kentucky belonged to the Daughters of the American Revolution, but their father (named David Foreman) was a mere private in the Confederate Army. I think everyone else in that august body was at least a captain.

One of my father's ancestors settled in Calvert County, Maryland, in the early 1600s. A century later, his descendants plowed new soil in Virginia's Shenandoah Valley. After two or three generations there, they followed Daniel Boone over the Wilderness Road into the dark and bloody ground of Kaintuck in the closing days of the American Revolution. My mother's family settled in New Hampshire before the Revolution, trekked to Ohio during the early 1800s, then to Missouri before the Civil War, and on to a dry-land pinto bean homestead in pre-statehood New Mexico. My gene pool, originally from England, Scotland, Ireland, Sweden, Holland, Germany, and France, composted in the mold and mast of the great Eastern forest, drank the wild waters of the Ohio and Mississippi, and was seasoned by the howl of the wolf in Kentucky and the thunder of Bison in Missouri. Folk songs and frontier yarns tell my ancestors' tale. I want to believe in their story, I feel my belly grow warm with the drinking of their legends, and yet . . .

And yet, dark memories lurk like fading photographs in a family album, like ancient melodies sung by a withered crone to a baby. . . .

Blankets given with a smile to the Indians of Massachusetts; blankets infested with smallpox . . .

Huge piles of Bison bones bleached white by the sun . . .

Wagons filled to overflowing with the corpses of Passenger Pigeons, ducks, and curlews, making their way to urban markets . . .

Twenty-five couples dancing on the raw stump of a redwood . . .

Five million cubic yards of concrete occluding the flow of the Colorado River. . . .

Those new beginnings for my ancestors meant abrupt, violent endings for countless other forms of life and cultures. Can we have a second chance at a new beginning?

More than 350 years ago, the Pilgrims stepped off the *Mayflower* into what they described as a "hideous and desolate wilderness."

Roderick Nash, in *Wilderness and the American Mind,*[1] quotes Edward Johnson, writing in 1654: "the admirable Acts of Christ" had transformed Boston from "hideous Thickets" where "Wolfes and Beares nurst up their young" into "streets full of Girles and Boys sporting up and downe."

In those "hideous Thickets" cloaking New England grew White Pines 220 feet tall. To the south and west stretched the most diverse deciduous forest on Earth. In those woods the massive crown of an American Chestnut might shade a quarter of an acre, and Tulip Poplars reached nearly as far toward the clouds as did the northern White Pines.

Coming from the tamed fields and picked-over woodlots of England, Cotton Mather and the Puritans were terrified by the dark, forbidding forest. Like the Spanish in Mexico a century earlier, they turned their backs on the opportunity to kindle a genuine beginning. Instead, they set about to transform the new into the old. Mather created a theology that is still with us, one that justified the reduction of the new continent and the decimation of its native humans. This frightening land, with its dangerous inhabitants, was obviously the haunt of Satan. He had established his Kingdom here, the Wolfes and Beares and Catamounts were his demons, and the savages his faithful flock. This theology inspired a religious crusade to conquer the wilderness, to wrest it out of the hands of Satan and return it to the fold of the godly.

Penetrating ever deeper into the great forest over the next two hundred years, the pioneers were convinced that the big woods had to be cleared—for two reasons, utilitarian and spiritual. First, it was necessary to open the forest to let in sunlight and grow crops; and, second, because the trees harbored savages, wild beasts, and Satanic notions, the wildwood had to be conquered in a spiritual sense as well.

From the beginning of our sojourn in North America, we failed

---

1. New Haven: Yale University Press, 1982. Anyone who wishes to understand the American relationship to wilderness must read Nash's landmark book.

to learn from the new land and from the people already living here. In that sense, the history of the United States of America is a story of spiritual failure, of missed opportunity to fashion a truly new society—for the first time, a civilization in harmony with the land. Instead, as Alfred Crosby argues in *Ecological Imperialism*,[2] we created a neo-Europe, transforming the ecosystems of North America into a copy of Europe. Not only did we bring our crops, domesticated animals, weeds, pests, and diseases; we carried with us the attitudes toward nature that eight thousand years of Western civilization had developed.

Puffed up with lofty piety, we assumed we had nothing to learn from the "Indians" and felt no compunction in exterminating them and appropriating the land. Because they worshiped Lucifer, we believed, it was not un-Christian to massacre them when the safe opportunity presented itself, or to give them blankets infected with pestilence when they were too formidable to fight. Because they did not have a concept of land ownership and obviously did not use the full potential of the rich Earth, the land was free and available for whoever put his sweat into it. After the reduction of the savages, when it was safe to consider them with more liberal attitudes, we still believed we had little to learn from them. Instead, it was our Christian duty to teach them to be white men and Christians.

So rolled the course of Manifest Destiny for almost four hundred years. We have possessed the land, tamed it, civilized it. We have forced it to meet our expectations, regardless of the results. Moving West beyond the Eastern Deciduous Forest, we plowed the Great Plains and created the Dust Bowl in two generations. Refusing to learn from that, we have now, two generations later, created the conditions for a potentially even more disastrous Dust Bowl.

Awash in a wealth of wood, the Puritans chortled that they could heat their humble homes warmer than could any English lord. Lumbermen have chased the American forest from the Atlantic to the Pacific over the millrace of these centuries, scalping it in New

2. Alfred W. Crosby, *Ecological Imperialism: The Biological Expansion of Europe, 900–1900* (Cambridge: Cambridge University Press, 1986).

England, stripping it bare in Pennsylvania, Wisconsin, and Michigan. All the while they have told us not to worry, the trees are without limit. The timber frontier leapt the plains, and now we stand on the Pacific shore, shipping the last of the ancient forest to Japan.

Because English lords surrounded their country manors with green lawns, in America, where there were no lords, Everyman surrounded his manor, no matter how modest, with a lordly lawn. Today, even on the arid high plains of Denver, Colorado, Everyman must have his lawn, and never mind that it means tunneling under the Continental Divide to bring the necessary water to keep it green.

These stories go on and on. Men and women with dirty fingernails built a new Europe in this fresh land, but it was an *idea* that led them to do it, the idea of the land as commodity—an idea as old as history, indeed, the idea that was the engine for civilization.

To be sure, some were lured by the wilderness. The unspoken anxiety on the frontier was the fear engendered by the white who scorned civilization for the siren call of wildness—who "went Indian." To "go Indian" was to descend into savagery and godlessness. The inhabitant of the edge of European society had to gird his or her loins against the temptations of the back of beyond. Civilization was the thin veneer separating us from the beasts. Those who voluntarily went native and sought the company of the red man were considered to be pitiful sinners. There was no greater horror on the frontier than to be captured and carried away into the primeval by Indian raiders. The highest duty for whites on the cusp of civilization was to rescue children and women from that unthinkable hell. The history of the westward movement is sprinkled with such tales—the Oatman girls, Veronica Ulbrick, Cynthia Ann Parker. . . . Incomprehensible to the rescuers were those cases in which the rescued white did not want to leave the Indians. Such events were more common than our written history tells.

Nevertheless, a few iconoclasts did encourage us to learn from the Indians, to adapt ourselves to the New World instead of transforming it in our image. Thomas Morton was deported back to England by the Puritans partly because of his respect for the Indians and his attempts to scout a different trail. Henry David Thoreau, after exploring the Maine woods in the company of a Penobscot guide, retired to Walden Pond to reflect on the American story. At a time when the cult of Manifest Destiny was at its peak, he had the temerity to oppose the conquest of northern Mexico by the United States, and to utter the greatest heresy in twelvescore years: "In wildness is the preservation of the world."

In the generation after Thoreau, John Muir turned his back on a career as a brilliant inventor and lit out for "the University of the Wilderness." Walking a thousand miles to the Gulf of Mexico, he caught passage on a freighter to San Francisco. He then spent ten years exploring the Sierra Nevada with little more than matches, a blanket, tea, and bread. He questioned America's march of empire and treatment of the land as no one had previously dared.

Not until 1890, when the United States Census Bureau officially announced the close of the frontier, did it become acceptable to reconsider the American experience. Interest in Native Americans surged around the turn of the century, and many prominent Americans despaired that the loss of a permanent frontier would mean the weakening of the traditional virtues of hardihood. Unfortunately, little of this macho breast-beating went deep enough to honestly question the essential rightness of the American relationship to the land, to ask if we had chosen the wrong fork of the path from the very beginning.

In the twentieth century, a number of thoughtful writers, ranging from farmer's wife Mary Austin and forest ranger Aldo Leopold to the English novelist and poet D. H. Lawrence, have poked and probed and worried at the thick knot of our seventeenth-century choice. John Muir, more than any other individual, had launched the nature preservation movement, and an undercurrent of skepticism continued to run through it, an undercurrent in the mainstream of conservation that bubbled up now and again with concern over that Puritan decision.

But an organized effort to find our way back to the rocky Plymouth shore, to locate through the fog of time the overgrown fork in the trail that we had missed so many lifetimes ago, had to await a maturing of both the environmental movement and the 1960s back-to-the-land movement. The poet Gary Snyder planted himself in the northern foothills of the Sierra Nevada in California and sank roots he hoped would endure for a thousand years. San Franciscans Peter Berg and Judy Goldhaft began to wonder what it meant to *dwell* on the Pacific Coast. In the hardscrabble Ozarks of Missouri, David Haenke mulled over how a sustainable culture could form in his place—not just learning from the old-timers and their ante-plastic/electricity/gasoline skills, but digging deeper to learn from the oak-hickory hills. From the Gulf of Maine to Inland Cascadia to the Sonoran Desert, a notion sparked and caught on dry tinder.

In 1981, idea entrepreneur Stewart Brand, of *Whole Earth Catalog* fame, asked Peter Berg and Stephanie Mills to guest-edit a special issue of his *CoEvolution Quarterly* magazine about bioregionalism—the name given to this confluence of second-chancing the American experience.

By 1984, bioregional groups had sprung up throughout the continent like fairy rings after a warm rain, and the first North American Bioregional Congress was called in the Ozarks.

One of the key concepts of bioregionalism is that modern political boundaries have no relationship to natural ecological provinces. Bioregionalists argue that human society—and therefore, politics and economics—should be based on natural ecosystems. They find affinity with Indian tribes and with Basque, Welsh, and Kurdish separatists, and have no sympathy with the modern nation-state, empire, or multinational corporation.

Peter Berg and Raymond Dasmann, in the 1978 book *Reinhabiting a Separate Country*,[3] charted the course of bioregionalism. Living-in-place, they say, "means following the necessities and pleasures of life as they are uniquely presented by a particular site, and evolving ways to ensure long-term occupancy of that site. A society which practices living-in-place keeps a balance with its region of support through links between human lives, other living things, and the processes of the planet—seasons, weather, water cycles—as re-

3. San Francisco: Planet Drum Foundation, 1978.

vealed by the place itself. It is the opposite of a society which *makes a living* through short-term destructive exploitation of land and life."

They further say, "*Reinhabitation* means learning to live-in-place in an area that has been disrupted and injured through past exploitation. It means becoming native to a place through becoming aware of the particular ecological relationships that operate within and around it."

Bioregionalism, then, is fundamentally concerned with dwelling in place, a concept far removed from the suburbs, cities, and farms of our continent. Reinhabitation involves adapting yourself to the place instead of the place to you; it means becoming part of a community already present—the natural community of beasts and birds and fish and plants and rivers and mountains and plains and sea. It means becoming part of the food chain, the water cycle, the *environment* of a particular natural region, instead of imposing a human-centered, technological order on the area. Along the North Pacific Coast, it means joining a community of salmon, Douglas-fir, big rivers, rain; in southern Arizona, a community of Saguaro, Javelina, ephemeral washes, summer thunderstorms; in the Northern Rockies, a community of Grizzly, Elk, Lodgepole Pine, long winters, and so on.

For us in America, bioregionalism means going back to that choice made by Cotton Mather and turning onto the other path. It means a second chance in the New World, another opportunity to meet that spiritual challenge.

While there are many currents within the bioregional movement, I hope most of us who consider ourselves part of it would agree on these points:

■ We are reaching for a new definition of "community," of belonging.

■ We reject the sacredness of progress and technology; instead we turn to craft, and to being.

■ We recognize that we are part of the natural ecosystem in which we dwell.

■ We are seeking new (old) ways of organizing ourselves, turning away from hierarchy to tribalism.

■ We dance, instead of marching.

■ We are subverting the dominant paradigm, not reforming it; and we subvert it by creating our own world, by avoiding a head-on confrontation—by using the might of the machine against itself in a political application of Eastern martial arts.

In these ways the bioregional movement goes beyond what have unfortunately become the narrow parameters of the environmental movement. Environmentalism has become a reformist but loyal courtier to the dominant industrial order. The worldview of environmentalism includes half a dozen billion human beings, nation-states, private automobiles, and people in business suits on every continent.

There is no hope for reform of the industrial empire. Modern society is a driverless hot rod without brakes, going ninety miles an hour down a dead-end street with a brick wall at the end. Bioregionalism is what is on the other side of that wall.

Of course, even for a collection of counterculture back-to-the-landers, cutting-edge post-liberal thinkers, and radical preservationists, it is not easy to question the entire European experience on this continent. While we want to be a product of this land—of Turtle Island, as the Indians named North America—we are a product of nearly four centuries of English colonization here, and of eight millennia of Western civilization.

It is hard to stand on the western rim of the Atlantic, with our backs to Europe, and realize that here can be something new, something pure, something clean, something right. It is difficult—with this second opportunity—to cease being English and become American. It is yet more difficult to do so when our memories of the New Land are faint, lost in the smoke and mist of history and myth, when the only North America we know well is the one recast in a European mold. These difficulties, together with the attractiveness of the easy life of this neo-Europe, lure us away from the rough trail in the swamps and White Pine groves of Massachusetts.

It is, then, little wonder that the bioregional movement seems to have bogged down in the mud and thickets of that trail, that the difficulty of the way sends us back to the easy, already-traveled one.

I discern three general areas where bioregionalism is encouraged to backtrack to the broad highway of American history or into dead ends of irrelevancy: an infatuation with *appropriate technology;* a thoroughgoing faith in *local control;* and an emphasis on *human beings.*

Appropriate, small-scale technology and simple, gentle, Earth-friendly life-styles are substantial aspects of bioregionalism. By developing such tools and ways of living, bioregionalism not only challenges the environmental movement to practice what it preaches, but it offers our global business society a sustainable, healthier alternative. Yet bioregionalism too frequently becomes mired in its composting toilets, organic gardens, bicycles, handicrafts, recycling, solar collectors, wind generators, barter systems, woodstoves. . . . These means of a sustainable life-style are important, yes, but bioregionalism is more than *technic;* it is resacralization and self-defense. Appropriate technology is a means to an end; it is not the end itself. Because the Yankee tinkerer and shade-tree mechanic are American folk heroes, this clever inventiveness becomes emphasized for its own sake instead of for potentially lessening our impact on the planet. The goal is reinhabitation; appropriate technology is a way to that goal, but it is not the goal.[4]

In its glorification of local control, bioregionalism may end up subverting itself to fit with the natural-world-as-supermarket mentality of the bumpkin proletariat of North America's rural areas. While local control of the land is fine in theory and as a long-term goal (after we truly appreciate this land and agree to adapt ourselves to it instead of the other way around), let us remember, for example,

4. Christoph Manes, author of *Green Rage,* and John Davis, editor of the *Earth First! Journal,* contend that no technology is appropriate, that technology is fundamentally different from tools and craft. Their point of view should not be dismissed as mere semantics; it may get to the heart of the matter of human manipulation of our environment.

that we would have little protected Wilderness or other natural areas in most of the Western states if it were up to the state-level politicians or rural residents of those states.

As a former conservation lobbyist, I can rave for hours about the ineptitude and stupidity of members of Congress, their lack of interest in things natural, and the control industrial corporations have over them. But Congress is a shining beacon of ecological enlightenment when compared to most state legislatures or, worse yet, to a rural county commission. We would not have 100 million acres of National Parks, Preserves, Monuments, Wildlife Refuges, Wildernesses, and Wild Rivers in Alaska if that issue had been left to the people and politicians of Alaska. The recent National Forest Wilderness bills for Utah, Wyoming, New Mexico, Arizona, and other states would not have passed if a *national* perspective and constituency had not been brought to bear on the issue. Mediocre as they are, these bills have saved some wilderness. And in the Eastern states with Wilderness bills, and in California, Oregon, Washington, and Colorado, you can bet your next hike that it was the urban dwellers and not the local country folk who carried the day for preservation.

During my own bucolic days in a small town in the Gila National Forest of New Mexico, I saw back-to-the-landers move to the area with wilderness dreams in their eyes and a willingness to act politically to protect the wilderness; but as the need for dollar bills and for social acceptance among the good ol' boys and gals of the county grew on them, some of these Ms. & Mr. Naturals became seedy rednecks complete with chain saws, trap lines, muscle wagons, and tight-lipped complaints about how the federal government was restricting their "right" to develop "their" natural resources for dollars in the bank.

Local versus national control of the public lands can be debated endlessly, but bioregionalists must take great care not to let an idealistic goal of local control and self-sufficiency destroy the higher goal of continuation of wild natural diversity.

For bioregionalism to last, to present a clear alternative, to create a world beyond the collapse of the industrial state, it must be concerned with deeper matters than alternative technology and non-

47

hierarchical human society—with more important matters than *human beings*. Bioregionalists must resist the scolding entreaties from the left and the right and the mainstream to be more concerned with social and economic issues than with ecological ones. The philosophies, worldviews, and religions of bioregionalism must be *biocentric*. If we are to reinhabit a place, form a community with the others that dwell therein, and have a second chance in America, then we must resist the temptation to put ourselves in the role of stewards. The other beings—four-legged, winged, six-legged, rooted, flowing—have just as much right to be in their place as we do; they have value completely apart from whatever worth they have for humans.

Of course, human inhabitants must make a living, too. Food, water, shelter, and other needs must be met. Advocates of Deep Ecology argue that human beings have the right to seek the fulfillment of their *vital* needs from an ecosystem. Problems arise when humans go beyond their vital needs or when they overshoot their carrying capacity. Paul and Anne Ehrlich point out that the impact of humans on a natural community is a product of population, technology, and affluence. Increase any of those factors, and the impact is increased. Determining how a biocentric philosophy is to be practically realized—how to limit our impact—will be a central task of bioregionalism for the next thousand years.

At its most basic, biocentrism can be manifested by reinhabitory humans in the same way that Oneida Indians kept a sense of their place in the natural world. According to an Oneida story,[5] the ancestors of the Oneida once grew in population so much that some of them had to go look for a new place to live. They found a wonderful place, and the people moved there. After moving, they found that they had "chosen the Center Place for a great community of Wolf." But the people did not wish to leave. After a while, the people decided that there was not room enough in this place for both them and Wolf. They held a council and decided that they could

5. Paula Underwood Spencer, *Who Speaks for Wolf* (Austin: Tribe of Two Press, 1983).

hunt all the wolves down so there would be no more. But when they thought of what kind of people they would then be, "it did not seem to them that they wanted to become such a people."

So the people devised a way of limiting their impact: In all of their decisions, they would ask, "Who speaks for Wolf?" and the interests of the nonhuman world would be considered.

In all our councils, in all our decisions, both individual and collective, we must not forget the others who are not represented. We must represent them ourselves. *Who speaks for Wolf? Orca? Gila Monster? Red-cockaded Woodpecker? Bog Lemming? Big Bluestem? Oak? Mycorrhizal fungi?* We must constantly extend the community to include all.

Further, we must demonstrate self-restraint and practice self-defense. In every bioregion there should be vast areas off limits to human use, simply left alone to carry on the important work of evolution.

The centerpiece of every bioregional group's platform should be a great core wilderness preserve where all the indigenous creatures are present and the natural flow is intact. Other wilderness preserves, both large and small, should be established and protected throughout the bioregion, and natural corridors established to allow for the free flow of genetic material between them and to such preserves in other bioregions. These preserves should not just include the "rock and ice" of high, remote mountains, but some of the gentler, biologically more productive land in every bioregion as well. The development of such proposals and the work to establish them should be key parts of bioregional activity. In many cases, temporary transitional management will be needed to help nature restore suitably large areas to wildness. Extirpated native animals should be introduced if possible. If salmon streams need to be repaired, clearcuts rehabilitated, prairies replanted, roads removed— then that becomes the hands-on work of reinhabitation.

These core wilderness preserves should be sacred shrines to us as reinhabitory people, but they transcend even their sacredness to us in being simply what they are—reserves of native diversity. Places beyond good and evil, places where being can simply be.

Although working toward such preserves is a central task of bioregionalism, it can be argued that the establishment of many such preserves is unlikely under the present human regime, and that their actual establishment will come on the other side of the wall. Nevertheless, they should form the center around which all of our other endeavors revolve from this day forward.

And that is where warrior societies like Earth First! come into the bioregional world. In reinhabiting a place, by dwelling in it, we become that place. We are *of* it. Our most fundamental duty is that of self-defense. We are the wilderness defending itself. By developing our own "land-use plans" with wilderness preserve cores and connecting corridors to carry the germ plasm of wildness, we chart the game plan for our own defense. We develop the management plan for our region. We then begin to implement it. If the dying industrial empire tries to invade our sacred preserves, we resist its incursions. In most cases we cannot confront it head-to-head because it is temporarily much more powerful than we are. But by using our guerrilla wits, we can use its own massed power against itself. Delay, resist, subvert, using all available tools: File appeals and lawsuits, encourage legislation—not to reform the system but to thwart it. Demonstrate, engage in nonviolent civil disobedience, monkeywrench. Defend. Deflect the thrashing mailed fist of the dying storm trooper of industrialism as represented by the corporate honcho, federal bureaucrat, and tobacco-chewing Bubba.

Our self-defense is damage control until the machine plows into that brick wall and industrial civilization self-destructs as it must.

Then the important work of bioregionalism begins.

5 | The Arrogance of Enlightenment

There are some who can live without wild things, and some who cannot.

—*Aldo Leopold*

I have heard it said that a wise man can find wilderness in a courtyard garden, can see a Grizzly Bear in a hothouse flower. Perhaps. Perhaps an enlightened woman or man can find such natural peace, such wildness in the mundane, such gladness in an artificial world. This kind of ability is undoubtedly healthy for modern people. It allows one to rise above the tawdry mess of civilization, to find unity with Nature even when that natural unity has been destroyed. It brings peace, contentment, and serenity. It prevents ulcers and high blood pressure.

But what does it do for the wilderness? What does it do for the Grizzly?

Where is the real world? What is reality? Is it within ourselves—in our minds, our consciousness? Is reality only what we perceive? Are our minds paramount, with no reality apart from our heads?

No! The real world is out there—independent, autonomous, sovereign, not ruled by human awareness. The real Grizzly is not in our heads; she is in the Big Outside—rooting, snuffling, roaming, living, *perceiving on her own*. Wilderness is not merely an attitude of mind; it is greater, far greater, than ourselves and our perceptions of it. We do not create reality; reality creates us. It is not "I think, therefore I am"; it is "I am, therefore I think."

These are important distinctions. Too many who seek enlightenment fail to see that the external reality of the redwood is more important than our attitude toward it, that the wilderness is greater than we, and thus greater than our illuminated view of it.

There are those, for example, who criticize secretive monkeywrenching for not being uplifting. They argue that if one destroys destructive machinery in order to defend the wilderness, one should do it in the open, in some kind of holy manner. Here the confusion is made complete: The results of protecting Earth are not important; it is the enlightenment or the uplifting from it that counts. The actual deed is unimportant; only the brownie points racked up in heaven are significant. What arrogant religiosity! What a mad delusion to think that one's mental gyrations are more important than the *reality* of actually protecting a two-thousand-year-old redwood or a hundred-ton Great Blue Whale or an unpeopled wilderness that is a nation unto itself!

I fear that those who argue for the process of action rather than the results are distracted by the ideas of an afterlife. Christian, Buddhist, New Age, whatever—those suffering from the arrogance of enlightenment are sometimes those who see Earth as a mere way station in the eternal progress of their souls.

If you want heaven—it is here. Walk through an aspen grove on a bright autumn day. The gold in that light is more real than in the streets beyond the Pearly Gates. If you seek total union with the cosmos, then float a river, drift into river time, let the rich red of the San Juan or the crystal of the Salmon make you part of All. If it's Valhalla you desire, stand with your bold friends before a bulldozer and then eat, drink, and make merry with them in victory celebration afterward. And reincarnation—yes, that, too. Your atoms are of the everlasting rocks, and will become buzzard, weasel, dung beetle, worm, and so on for eternity after your simple brain sleeps. Heaven, Nirvana, Valhalla, everlasting life are here and now—in the *real world*. We need nothing more than this paradise in which we were born.

The world exists independently of us. When a tree falls in the forest and no human is there to hear it, it still falls, the shock waves still echo from bluff to cliff, the bears and the birds yet hear, and life goes on. Only an arrogant fool could think otherwise. We can sit in perfect peace and contemplation in our manicured gardens, but if there are no Grizzlies in the Big Outside, there can be no Grizzlies in our flowers. And if, after the last redwood is cut, we are able to say, "Ah, but I had an enlightened appreciation of the essence of redwoodness," then our words will be the sound of one hand clapping. If we discourage others from acting boldly in defense of the wilderness because their hearts are not yet pure, then we become quislings to life.

Do not misunderstand my words. I seek after wisdom and enlightenment, too. I go alone into the wilderness in quest of visions. I sit in high, windy places and listen to the powers of the Earth. But I try not to delude myself with my own self-importance. I do not for a moment pretend that I am any more than an insignificant speck in this rich, voluptuous, living Earth. I try not to puff myself up so that I enthrone reality within my skull box. Reality is out there. In the Big Outside. And my action in defense of it—raw, rank, brawling, and boorish as it may be—is vastly more important than all the enlightenment with which I can swell my head in the several score years in which my consciousness exists.

6 | The Wilderness Gene

For me and for thousands of others with similar inclinations, the most important passion in life is the overpowering desire to escape periodically from the strangling clutch of mechanistic civilization. To us the enjoyment of solitude, complete independence, and the beauty of undefiled panoramas is absolutely essential to happiness.

—Bob Marshall

In the early 1970s, The Wilderness Society began a national series of personal development workshops in order to equip conservation activists with "people process skills." I remember sitting in a get-acquainted circle of fifty people at the Salt Lake City Howard Johnson's in 1973 while each of us related what had brought her or him into conservation activism. Apart from the heartfelt (and inspirational) passion revealed in each person's statement, what struck me was that there was no single theme uniting our commitment to wilderness. For some, a particular incident in childhood was remembered; for others, there had always been a connection with "things wild and free." Some had parents who took them to the Big Outside, others had to find that path by themselves. Some had been struck as Saul was struck on the road to Damascus; for others, there never had been any sense of being "reborn" or awakened.

In dozens of similar conferences, all with the same opening session, this lack of pattern was repeated. It was noticed and became a topic of lively conversation between me and a few other junior members of The Wilderness Society staff (we had endured many of these workshops). What made us wilderness nuts different? Why did most people care less? If there was some common source of enlightenment, a certain factor we had all experienced, a particular trigger to pull, then we only needed to duplicate it for the mass of humanity and everyone would want to preserve the wilderness!

Sadly, we could discern no common strand that pulled wilder-freaks to the wild. The diversity of wilderness activists in terms of life-style, personality, work, location, age, background, income, politics, and philosophy was mirrored in the reasons why they worked for wilderness preservation.

Finally, one of my colleagues half-jokingly suggested the "wilderness gene." Maybe attachment to the wilderness is determined not by one's experiences but, rather, genetically—like eye color.

Over the years, as I've chewed on his idea late at night, I've accepted it more and more, and I've speculated even further: Maybe this wilderness gene reaches far, far back into our evolutionary history as *Homo sapiens*, back to the first of our species—the Neanderthal. Unlike our own bull-in-the-china-shop kind, there is no evidence that Neanderthals ever got out of balance, ever upset their environment, ever forgot their place in nature, ever caused the extinction of other species. (It was modern humans who were involved in the extinction of the Pleistocene megafauna twelve thousand years ago.)

Perhaps, as Cro-Magnons moved into Neanderthal territory, displacing them and causing their extinction, Neanderthal genes were picked up thirty millennia ago by the Cro-Magnon gene pool (through tried and true techniques), thereupon to drift along beneath the surface, bubbling up now and then in a Lao Tzu, a Saint Francis, a Dogen, a Mary Wollstonecraft, a Chief Seattle, a Thoreau, a Muir, a Mary Austin, a Rachel Carson.

To approach this midnight tachyphrenia from another angle, let's consider the role we modern humans play in the overall ecology of our planet. If we accept the Gaia hypothesis that Earth functions as a single organism, then where do we fit in? The modern bearers of Classical Greek rationalism and hubris proclaim that we are Gaia's nervous system—the brain, the communications aspect, that we are ensconced in the driver's seat. Wiser students, such as the eminent University of Chicago historian William H. McNeill, instead compare us to a disease:

> *Looked at from the point of view of other organisms, human-kind . . . resembles an acute epidemic disease, whose occasional lapses into less virulent forms of behavior have never yet sufficed to permit any really stable, chronic relationship to establish itself.* [1]

Admittedly, this characterization sticks in one's craw. If there's one thing upon which all "civilized" religions and philosophies agree, and have agreed since the first civilization crawled out of a Sumerian irrigation ditch, it's that human beings are the top dogs on the planet. It runs entirely against all that's been funneled into our heads since birth to see ourselves as a disease. Yet if we attempt to approach a viewpoint of biological objectivity, if we try to ponder things even briefly and imperfectly through the eyes of other species, the truth hammers at us relentlessly: In our decimation of biological diversity, in our production of toxins, in our attack on the basic life-support system of Earth, in our explosive population growth, we humans have become a disease—the Humanpox.

The bodies of individual creatures, when confronted by disease or invasive organisms, produce antibodies and send out the marines—white blood cells (phagocytes), who, without regard for their own welfare but only for the sake of the greater body, fight, consume, and overpower the invading disease organisms.

Perhaps this is what has happened with conservationists. As the Humanpox has metastasized from a simple, uncomfortable, localized skin rash to a systemic life-threat, Gaia has reached into the disease itself for antibodies. That long-buried Neanderthal gene has been pulled to the surface, and in grim retribution for the slaughter of Neanderthals by modern humans (Cro-Magnons), a new race of

---

1. William H. McNeill, *Plagues and Peoples* (Garden City, N.Y.: Anchor/Doubleday, 1976).

Neanderthals, humans who love the wild, whose primary loyalty is to Earth and not to *Homo sapiens,* have been born, and they will fight like antibodies and phagocytes for the wild, for the precious native diversity of our planet.

A wild flight of fancy? Perhaps. Unscientific? Thoroughly. But I think about it on stormy nights, while sipping yesterday's wine and smoking cigars, when the moon is "a ghostly galleon tossed upon cloudy seas." I mull it over when the cup is in my hand, the wind in my face, the smell of rain in my nostrils, the raw power of nature brooding overhead with lightning and ragged clouds. Antibodies. Antibodies against the Humanpox. The revenge of the Neanderthal. The Wilderness Gene.

Antibodies need no justification. Their job is merely to fight and destroy that which would destroy the greater body of which they are a part, for which they form the warrior society.

Guess I better ooze on out of here and take care of that bulldozer. . . .

7 | Preserving the Wilderness Experience

Of what value are forty freedoms without a blank spot on the map?

—*Aldo Leopold*

rowing up in Albuquerque, New Mexico, the blank spot on my map was the Gila Wilderness, two hundred miles to the southwest. As I explored nearby wilderness such as the Sandia, Jemez, and Sangre de Cristo mountains, the Gila took on mythic proportions in my mind. It was vast, remote, incomparably wild . . . and unknown. It became my El Dorado, my ultima Thule. It was the Heart of Darkness, the center of magic.

As soon as I was able, I began to hike the trails, climb the summits, and explore the canyons of this thousand-square-mile mountain fastness. In a quarter-century of enjoying the Gila, I've been flooded out of campsites by torrential rains, blasted off ridges by lightning, and run out of the high country by late and early snowstorms. Once I lost a food cache to a bear. Back in the early seventies, when I was running a part-time guide service to supplement my meager Wilderness Society salary, I lost two mules in that Big Outside; on one of those occasions, I had to pack my wife out on my back after her knee was wrecked by her mule, Nellie Belle.

Even with misadventures such as these in the Gila, and those elsewhere, ranging from a summer blizzard in Alaska's Denali to a rockslide above camp in Chihuahua's Barranca del Cobre, I gradually grew fat and careless from an abundance of easy days and nights in the backcountry, allowing lazy familiarity with wilderness travel to dull my primitive skills and alertness.

A trip in the Gila in August 1988 gave me my comeuppance and put back the edge. My wife (a different one from the mule victim) and I invited my twelve-year-old nephew, Gerard, on a ten-day backpack into the Gila Wilderness. We planned a soft trip—Nancy and I, while healthy, were typically out of shape, our ten-day packs felt like past sins riding our shoulders (my pack started out heavier than my nephew), and, because it was only Gerard's second backpack, we didn't want to take a killer slog that would scare him away from future trips.

Our destination was McKenna Park, a ten-square-mile plateau bounded by the rimrocked canyon of the West Fork of the Gila River on the north, the Diablo Mountains on the east and south, and the higher Mogollon Mountains on the west. Here was the farthest point from a road in New Mexico, and the most pristine Ponderosa Pine forest in the Southwest.

McKenna Park has even been protected from livestock in recent decades. The native Merriam's Elk had been exterminated early in the century by commercial hunting to feed mining camps, so the New Mexico Game and Fish Department introduced Elk from Yellowstone to take their place in 1954. To ensure healthy habitat, cattle had been removed from much of the Gila Wilderness, including McKenna Park. Thanks to this thirty-year respite from the tireless jaws of domestic bovines, McKenna Park looks and functions as a Ponderosa Pine forest is supposed to look and function. Beneath the 150-foot-tall pines and fifty-foot-tall Gambel Oaks is a robust carpet of native grasses and Bracken Fern.

Our trip into McKenna was not difficult, although August is thunderstorm season in the Gila and we had to tramp through daily downpours and dodge lightning over the Diablo passes. Hiking through the ancient forest, we spooked a Great Horned Owl out of a grandfather pine and stumbled onto a herd of Elk lazily grazing in bunchgrass belly-high. Gerard and Nancy's birding competition heated up: Lark Sparrow, Painted Redstart, Dusky Flycatcher, Red Crossbill, Western Wood Pewee . . . enlarging their life lists.

We planned to return to our truck at Gila Cliff Dwellings National Monument by hiking down the West Fork of the Gila River. I had walked that route on half a dozen previous occasions. Although there were dozens of fords in the sixteen miles, I remembered the stream as placid and no more than mid-calf deep. We planned to take three days to hike out, which would leave us unhurried afternoons to fly-fish the West Fork.

The wilderness had different plans.

Because of the unusually heavy rainfall that summer, the West Fork was roaring. The rushing water was above my knees, but the first crossing wasn't too difficult. Armed with alder hiking sticks, Gerard and Nancy picked their way across the ford. We realized this would not be a leisurely stroll down the canyon, but reckoned it would be passable without mortal danger. As we forged our way downstream, though, the current ran faster, the crossings grew deeper, the trail became a long, muddy puddle, and the rain continued to pelt us. Soon the river had climbed to my waist and was trying to strong-arm me downstream. I moved my sleeping bag from its usual place at the bottom of my pack to the top. Nancy—who had once lost her backpack and been stranded on a rock midriver in Northern California's Ishi Wilderness after being tossed off her feet by a snow-melt surge—gritted her teeth and fought her way across the fords. But Gerard, a short little squirt, couldn't wade the torrent without getting his entire pack (and sleeping bag) soaked. This meant I had to make three trips through each ford: one for my Kelty, one back, and then a third piggybacking Gerard and his pack.

To make our requisite five miles a day, we had to trudge most of the daylight hours. Our enthusiasm for bird-watching and fly-fishing washed out of us with the crossings; everything we had was going into making our way across those damn fords and then walking fast enough down the soggy trail through the continuous thunderstorm to stave off hypothermia. It was senseless to try to stay dry. Lightning smacked the rim and rock pinnacles above us. Pour-offs usually dry turned into cascading waterfalls. As the West Fork pounded me into humility, Gerard developed an admirable aptitude for fording the river. As I grew weaker, he grew stronger. Transferring his sleeping bag to the drier refuge of my pack, he half-skipped and half-swam across the current with the aid of his alder staff. Good thing, too. By the third day, I no longer had the stamina to muscle my way three times across each ford. Once, when old wilderness hands Nancy and Dave were at a loss as to how to get across a ford where the current ripped under a fallen cottonwood and created a potentially fatal situation, Gerard found a safe crossing for us upstream.

Late on the third day, as we hobbled into the parking lot, I smiled at my nephew and teased him, "I'll bet you can't wait to get back to Big Macs, video games, and TV."

Dripping like an alley cat that had been flushed down a sewer, he looked at me, looked back up the rocky canyon of the West Fork to a black-bellied thunderstorm growling overhead, and said, "Let's go back."

We were hurtin' buckaroos when we limped, cold, exhausted, and bleeding, out of the Gila Wilderness on August 24, 1988, but we were better for the challenge the stream had thrown at us. Gerard will remember that trip for the rest of his life; it will be a standard against which he'll measure all of his future trips into the back of beyond. Being able to handle the difficulties of the flooding West Fork at forty-one years of age made me feel pretty good. I wasn't as old or soft as I sometimes pretended. And Nancy was glad for the opportunity to test herself against another flooding wilderness river, and find she had the grit for it.

What we experienced in the Gila Wilderness was the tonic that draws many of us to the wilderness, the challenge that leads us to cast off gasoline-powered transportation for our own two feet, to decline a warm indoor bed for hard ground and a rain fly. What we too often find in Wilderness Areas, however, are not primitive conditions where we are on our own, but backpacking parks designed and managed for our comfort and safety.

It has been nearly seventy years since Aldo Leopold succeeded in having the headwaters of the Gila River in southwestern New Mexico protected as America's (and the world's) first Wilderness Area. The early advocates of wilderness preservation initially saw Wilderness Areas as places for quality (nonmotorized) outdoor recreation; for hunting, fishing, and pack trips; as opportunities to relive the pioneer experience; as testing grounds for the traditional American virtues of self-reliance and toughness. They believed, with George S. Evans, writing in 1904 (admittedly in sexist terms, but it was 1904), that "the wilderness will take hold of you. It will give you good red blood. It will make you a man."

In *A Sand County Almanac,* Aldo Leopold wrote:

> *Wilderness areas are . . . a series of sanctuaries for the primitive arts of wilderness travel, especially canoeing and packing.*
>
> *I suppose some will wish to debate whether it is important to keep these primitive arts alive. I shall not debate it. Either you know it in your bones, or you are very, very old.*

Many conservationists and biologists recognize today that the primary value of wilderness is not as a proving ground for young Huck Finns and Annie Oakleys. It is to preserve native biological diversity, to allow room on this human-dominated Earth for the free play of natural forces, to leave things alone *somewhere.* Nevertheless, preserving a quality wilderness experience for the human visitor, letting her or him flex Paleolithic muscles or seek visions, remains a tremendously important secondary purpose.

The agencies in charge of our Wilderness Areas have not sufficiently protected the recreational values of Wilderness Areas. (I discuss elsewhere the failure to designate and manage adequate areas for the preservation of ecological wilderness.) How might the Forest Service and other agencies manage already-designated Wilderness Areas in such a way as to provide a maximum challenge for the wilderness visitor who desires it?

The 1964 Wilderness Act permits a wide array of developments and artificial aids for the convenience of visitors to Wilderness Areas. These include constructed and maintained foot and pack trails, trail signs, detailed maps and brochures, the use of motorized equipment for search and rescue, and even permanent fire rings, primitive privies, boat docks, culverts, bridges, and spring developments where deemed necessary. Most hikers and packers want and expect some of these artificial aids. It is clear that in certain popular or especially delicate areas, well-engineered trails, designated campsites with fire rings, and sometimes privies are necessary to limit human impact on the wilderness ecosystem.

The failure of the Forest Service and Park Service is in not recognizing that a diversity of opportunity is needed. Some of us sometimes want to test our bodies and minds against *real* wilderness, and are offended even by infrequently maintained foot trails and rustic signs. Not all Wilderness Areas should be managed for the "gentle" backcountry experience. Not all parts of individual Wilderness Areas should be managed to provide equal ease of travel or safety. What is needed is a range of management options—from good trails and all the trimmings in places like Yosemite and the Sandia Mountains, to something approximating pre-European America in more remote and less-frequented areas. The wild end of the spectrum would be true wilderness where a hiker is utterly on his or her own.

A management scheme for these "primeval" areas would include several bold rules:

No trails. No new trails would be constructed. Existing trails would receive no further maintenance and would be allowed to deteriorate. If necessary to prevent erosion, certain trails would be put to bed. Routes through the Wilderness Area would be kept open only by the passage of visitors or wildlife. New blazing would be prohibited.

No signs. No trail signs, mileage markers, or location signs would be allowed anywhere in the Wilderness. At entry points the only sign permitted would be one identifying the Wilderness boundary and explaining that the visitor was entering a Primeval Wilderness.

No facilities. No boat docks, hardened campsites, hitching racks, and the like would be permitted. Fire lookout towers, administrative cabins, fences, and such would be removed or allowed to deteriorate. Visitors would be strictly required to practice "invisible camping" (obliterating fire rings, no cutting of vegetation, no ditching around tents, no littering, etc.)—which, of course, should be practiced in *all* backcountry areas.

No maps. The managing agency would print no maps or brochures for the area. The U.S. Geological Survey would discontinue printing and distributing topographic maps for the area. On maps covering a larger area (i.e., a map of a National Forest), the portion within the Primeval Wilderness would be blank; even topographical features (streams, ridges, peaks) would not be shown. Visitors would be discouraged from carrying old maps of the area with them. In very densely forested country, rough hand-drawn maps could be permitted. Detailed topographic maps encourage us to forget how to find a route with only a compass (if that) and our own wilderness sense.

No guides. Commercial guide and outfitting services would not be permitted. Visitors would have to go in on their own and confront the wilderness with only their skills and knowledge.

Hunting. Hunting would be permitted, if at all, only with primitive weapons (bow and arrow, atlatl, knife, sharp rock). Fly-fishing would be the only method of fishing allowed. Artificial stocking of fish would not be permitted.

No rescue. A visitor to a Primeval Wilderness would have to rely on her or his own resources for survival. No organized or mechanized search-and-rescue operations would be allowed in a Primeval Wilderness. If you break a leg, you crawl out, have your friends tote you out on their backs, or die. Mountain man John Colter didn't have helicopters to pull him out of Blackfoot country in 1808. He walked naked 150 miles back to Lisa's Fort.

Such a rescue policy would require federal legislation absolving managing agencies of responsibility for accidents. This kind of "hold harmless" legislation for visitors and users of federal lands is long overdue for *all public lands.* The fear of being sued by morons who harass and then are injured by bears, Bison, or Moose is what leads Yellowstone National Park to execute "problem" wildlife. No visitor to a National Park, National Forest, or other unit of public land should expect to be protected from the inherent dangers of the natural world. No damage suits should be allowed against agencies for not warning a tourist that a one-ton Moose is dangerous.

No modern equipment. In a few of these areas, even modern backpacking equipment (stoves, synthetic materials, aluminum pack frames, etc.) should be prohibited. At the very least, brightly colored tents and backpacks would be discouraged in Primeval Wilderness Areas.

Most areas in the National Wilderness Preservation System need not be managed in this way. As I meander through middle-age with the bad joints and beer belly that come with the territory, I enjoy good trails and topo maps as much as any hiker does. On a recent five-day trip over steep, narrow paths covered with deadfall in California's Ventana Wilderness, I ached for well-maintained trails. Nevertheless, sometimes you need challenge instead of recreation. For those whose blood thirsts for Evans's tonic, a few Wilderness Areas should be managed to provide that ultimate Primeval Wilderness experience. Among the areas that might be ideal for this type of management are the Aldo Leopold Wilderness in New Mexico's Gila National Forest, the Cabeza Prieta National Wildlife Refuge in Arizona, the Dead Horse Mountains/ Boquillas Canyon section of Texas's Big Bend National Park, the Mazatzal Wilderness in Arizona's Tonto National Forest, the Desert Game Range in Nevada, the northeast corner of Yellowstone National Park and the North Absaroka Wilderness (Shoshone National Forest) in Wyoming, the Maze District of Utah's Canyonlands National Park, and the Five Ponds Wilderness in New York's Adirondack State Park. These areas attract relatively few visitors, their trails are generally in poor condition, and they are very wild. This kind of challenge need not be limited to large, remote areas. There are other Wilderness Areas—such as Ishi in California, Cruces Basin in New Mexico, West Clear Creek in Arizona, Box-Death Hollow in Utah, and some in the East—that are small but could be managed without trails, maps, the promise of rescue, or other conveniences for visitors.

Many areas that the Bureau of Land Management is studying for Wilderness designation would be suitable for Primeval Wilderness management. These include the Black Rock Desert in Nevada, Alvord Desert in Oregon, Dirty Devil River in Utah, Arrastra Mountain in Arizona, Saline Range in California, West Potrillo Mountains in New Mexico, and Great Rift in Idaho. These BLM areas are essentially without constructed trails or visitor facilities. There should also be Wild Rivers where the river runner could experience this same degree of adventure. Possible candidates are the Lower Canyons of the Rio Grande in Texas, the upper Owyhee in Idaho and Oregon, the Verde in Arizona, and the Illinois in Oregon. Several of the sprawling, remote Wilderness and Wild Rivers in Alaska are obvious possibilities for Primeval management. Finally, portions of some large Wilderness Areas, such as the River of No Return in Idaho, Boundary Waters in Minnesota, Grand Canyon in Arizona, and Everglades in Florida could be managed as Primeval Wilderness while the rest of the area was managed with no limitations on Wilderness Act–permitted improvements for the visitor.

Although I'm focusing here on recreation and aesthetics, the greatest advantage of Primeval management would be to benefit wildlife by limiting the number of people in such areas. Nonetheless, there would need to be careful monitoring of these areas; if signs of excessive human impact (erosion from poorly located travel routes, for example) appeared, then limits would need to be placed on the number of people permitted to use the area, or carefully planned trails might need to be built.

A step beyond Primeval management would be human exclosure zones: large areas where no human beings, including scientific researchers or rangers, would be permitted. But that is another topic.

In 1930, Bob Marshall wrote that wilderness should provide an "opportunity for complete self-sufficiency." It's high time we put that philosophy into effect, and made room for a few blank spots on the map.

8 | The Destruction of Wilderness

The universe of the wilderness is disappearing like a snowbank on a south-facing slope on a warm June day.

—Robert Marshall

Humans have been destroying wilderness for millennia. Before agriculture was midwifed in the Middle East, humans were in the wilderness. We had no concept of "wilderness" because everything was wilderness and *we were a part of it.* But with irrigation ditches, crop surpluses, and permanent villages, we became *apart from* the natural world and substituted our fields, habitations, temples, and storehouses. Between the wilderness that created us and the civilization created by us grew an ever-widening rift.

Today, wilderness is destroyed in myriad ways. In fairness, we must recognize that *all* of us are destroying wilderness because of the alienation of our society from nature; because of human arrogance; and because of the gross overpopulation of our species combined with the wasteful life-style of modern humans, which converts 30 percent of Earth's photosynthetic production to human purposes. Living in the United States today entails consuming foods, paper products, fuels, and other materials that reach us at the expense of wilderness.

In this chapter I want to delineate in more detail the ways we are destroying wilderness in the United States of America in the late twentieth century. As we discuss this, we will also uncover *who* is destroying wilderness.

Road building. The army of wilderness destruction travels by road. With few exceptions, all attacks on wilderness require roads or motorized vehicles. The road network in the United States is pervasive: twenty-one miles is the farthest point from a road in the lower forty-eight states; there are few places ten miles or more from a road.

The National Forest System contains a large share of the Big Outside in the lower forty-eight states, but it also boasts 375,000 miles of road—the largest road network managed by any single entity in the world. The United States Forest Service employs the second-highest number of road engineers of any agency in the world (over 1,000). During the next half-century, the Forest Service plans to build an additional 350,000 to 580,000 miles of road, mostly for logging. At least 100,000 miles of that will be in currently roadless areas. This road construction costs the American taxpayer half a billion dollars a year. Reducing—or, better yet, eliminating— the bloated Forest Service road building budget in the congressional appropriations process is one of the best ways to defend wilderness. Simply writing one's members of Congress and demanding that the Forest Service road budget be cut or eliminated is one of the more effective single acts any of us can perform.

The Bureau of Land Management is also beefing up its road network—for the benefit of graziers, energy companies, and motorized recreationists. Many BLM areas could be classified and protected as vast Wilderness preserves if one or two dirt roads were closed.

The wilderness of our best-known National Parks has been rent by "scenic motorways." The Going-to-the-Sun road in Glacier, Tioga Pass road in Yosemite, Skyline Drive in Shenandoah, Newfound Gap road in Great Smoky Mountains, Trail Ridge road in Rocky Mountain, and Island in the Sky road in Canyonlands are prime examples. A battle is now raging in Capitol Reef National Park in Utah as local boosters persevere in their effort to pave the Burr Trail.

Roads are used for logging; dam building; oil and gas exploration; overgrazing "management"; powerline construction and maintenance; mineral exploration and extraction; and ski area, recreational, and subdivision development. Trappers, poachers, slob hunters, prospectors, seismographic crews, archaeological site vandals, and other vanguards of the industrial spoliation of the wild use roads. Roads provide freebooters with access to key areas of wildlife habitat and to the core of wild areas. Roads cause erosion, disrupt wildlife migration, and create an "edge effect" that allows common weedy species of plants and animals to invade pristine areas that provide refuge to sensitive and rare native species. Many creatures are killed by vehicles on roads.

Without roads, without mechanized access, native species are more secure from harassment and habitat destruction, and fewer people with fewer "tools" are able to abuse the land.

Logging. As the pioneers encountered the frontier in their march from the Atlantic seaboard to the Mississippi River, their first step in civilizing the land was to "open it up": The oppressive forest, harboring savages, wild beasts, and godlessness, and shutting out sunlight and progress, had to be cleared. While much of this ancient forest was simply burned, some of it fed the growing timber industry, which quickly became dominated by larger and larger companies as the timber frontier moved from New England to the Upper Midwest to the Pacific Northwest. In the view of the timbermen, the forests were endless, and they felt perfectly justified in ransacking an area, leaving it raw and bleeding, and moving farther west. In the late 1880s and 1890s, public outcry over this rapaciousness led to the protection of Adirondack State Park in New York and the establishment of forest reserves in the West to protect watersheds.

John Muir hoped the forest reserves would be off limits to logging, but under the leadership of Gifford Pinchot, they became the

National Forests and were dedicated to "wise use." The early Forest Service hoped to sell its timber to private companies, but these companies still had plenty of old growth on their millions of acres of private lands, and were not interested. Not until after World War II did the marketing of National Forest timber attract interest as the stocks on corporate lands became depleted. In the last forty years, the annual cut on the National Forests has increased steadily, until today the Forest Service brags that it is logging (i.e., destroying) a million acres of wilderness a year.

It is important to keep in mind not only that "harvesting" 10–12 billion board feet of timber a year from the National Forests (about a fifth of the nation's total timber production) exceeds sustained yield (the amount of timber harvested is more than that grown), but also that most timber sales in remaining roadless areas on the National Forests are *below cost* sales. It costs the Forest Service (and thus the taxpayer) more to offer and prepare these sales for cutting than timber companies pay for them. The Office of Management and Budget reported that, in 1985, Forest Service below-cost sales cost the taxpayer $600 million. Moreover, this figure does not include the associated costs of destroyed watersheds, devastated wildlife habitat, loss of recreation, herbicide pollution of air and water, decreased native diversity, concentration of wealth in fewer hands, and bureaucratic growth in the Forest Service to administer the program.

The situation is getting even worse. According to a recent study by The Wilderness Society, forest plans nationwide call for an increase in logging of 25 percent over the next decade. Virtually every unprotected large, forested, roadless area in the National Forests is threatened with logging and associated road building. Except for the small amounts of old-growth forest in designated Wilderness Areas, the Forest Service plans to convert the remaining old growth to intensively managed tree farms during the next fifty years. (And the Forest Service's attempts at tree farming are not always successful: Many clearcuts have not regenerated, even with expensive replanting, fertilizing, and herbiciding. Hundreds of Forest Service clearcuts remain butchered, bleeding wastelands decades after logging.)

Grazing. The livestock industry has probably done more basic ecological damage to the Western United States than has any other single agent. The Gray Wolf and Grizzly have been exterminated throughout most of the West for the benefit of stockmen. Grizzlies are still being killed around Yellowstone National Park and the Rocky Mountain Front for sheep ranchers; the new Gray Wolf pack in Glacier National Park has been largely wiped out to protect cattle; and ranchers are the leading opponents of wolf reintroduction in Yellowstone and the Southwest. The Mountain Lion, Bobcat, Black Bear, Coyote, Golden Eagle, and Common Raven have been relentlessly shot, trapped, and poisoned by and for ranchers until Mountain Lion and Bobcat populations are fractions of their former numbers. Elk, Bighorn, Pronghorn, and Bison populations have been tragically reduced through the impacts of livestock grazing. Streams and riparian vegetation have been degraded almost to the point of no return throughout much of the West. The grazing of cattle and sheep has dramatically altered native vegetation communities and has led to the introduction of non-native grasses palatable only to domestic livestock. Sheet and gully erosion from overgrazing have swept away most of the topsoil in the West. In non-timbered areas, most "developments" on public lands—roads, fences, juniper chainings, windmills, pipelines, stock tanks, and the like—benefit only a few ranchers.

Mining. Although mining has affected a smaller acreage than have logging or grazing, where it has occurred, its impact has been momentous, as a glance at the Santa Rita open-pit copper mine in New Mexico, or uranium tailings around Moab, Utah, will attest. Besides the scarification of the land and attendant air, water, and soil pollution, mining requires a network of roads, power lines, pipelines, and other infrastructure that drive away wildlife and dispel wildness. Geological processes are such that minerals tend to be most concentrated in rugged terrain, which is not only more vulnerable to damage but is also more likely to be wild and roadless than is gentler country.

Mining on the National Forests and BLM lands is sanctioned by the 1872 Mining Act, an antique from the days of the early gold rushes in the Wild West. This law allows any individual or corporation to claim minerals on federal lands. Such claims are staked by only a small filing fee and maintained by only one hundred dollars' worth of work a year, and can be taken to patent (passed into private ownership) if a reasonable mineral production is made. Like logging and grazing, mining on the public lands is a gigantic rip-off. Most National Parks and Wildlife Refuges are closed to new claims, as were Wilderness Areas after 1984 (previous claims—those filed prior to 1984 for Wilderness Areas, or filed prior to Park or Refuge designation—can be mined in all of these areas, however). The Forest Service and the BLM are limited in restricting or regulating mining on their lands, although they have more authority than they exercise.

There are essentially two types of miners operating today: the "small miner" and the mining corporation. Small miners are typically ne'er-do-wells with a bulldozer and a fanatical conviction that they're going to make a big strike that they can sell to a large corporation for millions. These folks live in backwater towns near their diggings, or commute on weekends from Phoenix, Los Angeles, and other cities to the backcountry. Although these little guys have made virtually no large strikes, they seem to be everywhere in the West, and can be enormously destructive to wild country as they prospect. They are also likely to poach, trap, or pursue other unsavory habits. They are vocal and potentially violent opponents of Wilderness designation and other "lockups."

Medium-to-large corporations do the real mining. They have professional geologists, use sophisticated methods to locate potential ore bodies, and carry large exploration budgets. Although financially and institutionally better able to practice mining and reclamation in a less environmentally destructive manner than small miners, they are not inclined to do so unless forced. Mining companies have considerable political clout in the Western states, and they and their lobbying association, the American Mining Congress, are powerful opponents of Wilderness and National Park designations, arguing that all the public lands must remain available for more sophisticated prospecting techniques that will be developed in the future, so that they can patriotically produce the strategic minerals America needs to hold the worldwide godless communist conspiracy at bay.

A national effort to replace the 1872 Mining Law with a lease and royalty system that would include environmental safeguards failed in the late 1970s, owing to pressure from both types of miners. National conservation groups are again considering such a campaign. It is long overdue. Even more overdue is a ban on mining in all remaining wild (roadless) areas.

Energy extraction. Unlike hard-rock mining, energy extraction (oil and gas, coal, tar sands, geothermal) on the public lands by private companies is governed by leasing. Leasing, in contrast to claiming, returns fees to the federal treasury, and does not transfer ownership of the land from the federal government. It is based on several laws more recent than the 1872 Mining Law. Although the Secretary of the Interior has considerable discretion in leasing, the federal government (especially under the Reaganauts) has been enthusiastic to lease as much of its land as possible to the few giant corporations (Exxon, Mobil, Shell, Chevron, Union, Getty, etc.) that dominate all facets of the industry.

Exploration for oil and gas begins with seismographic crews, who use explosives or "thumper trucks" to produce vibrations in the ground. Subterranean echoes are then read on monitors to determine where potentially favorable geological formations exist. Because competing companies carefully guard their information, sometimes a dozen different seismo crews go over the same terrain. Their blasting disturbs wildlife, and thousands of miles of road have been bladed through Western wildlands for thumper trucks.

After a favorable formation is found and an exploration lease obtained, exploratory drilling begins. Roads are built into wild areas, drilling pads are cleared, and outsized drilling rigs are set up for several weeks or months. The roughnecks who work on such crews are often ORVers, poachers, pot hunters, and other unenlightened users of the wild. Even if a strike is not made, exploration roads frequently become part of the permanent road system of the National Forest or BLM district, and provide access to wild country for the motor-bound public.

If a strike is made, more wells are drilled, roads built, pipelines constructed, and pumping stations installed, until dozens of square miles of public land become an industrial complex, and Elk, bear, and other critters are displaced. Such is the scenario for hundreds of thousands of acres of roadless country in the so-called Overthrust Belt of the Central and Northern Rockies.

Geothermal leasing, exploration, and extraction generally follow the same pattern as that for oil and gas. Coal, usually strip mined, and tar sands are leasable minerals on the public lands. They are a threat to wilderness primarily in Utah.

Dams and other water developments. Some of the most remarkable wildlands and rivers in the United States have been flooded by dams and their reservoirs. Glen Canyon (on the Colorado River in Arizona and Utah), Hetch Hetchy (in Yosemite National Park, California), and much of Hells Canyon (Snake River on the Idaho-Oregon border) have been drowned beneath stagnant reservoir water. Only all-out national campaigns by conservationists have prevented dams in the Grand Canyon, Dinosaur National Monument, the Gila Wilderness, and the remainder of Hells Canyon. Dams on the Columbia River have decimated salmon runs in the wilderness of the Northwest and Central Idaho. Upstream dams on the Colorado, Green, and Rio Grande have severely affected wildlands downstream.

These dams have been built by the Army Corps of Engineers, the Bureau of Reclamation, and the Bonneville Power Administration for electric power generation, flood control, irrigation, and "recreation."

The era of giant dam building in the United States is coming to a close, and only a few large roadless areas are threatened by future construction. A new threat, however, is that of "small hydro," the construction of small dams and power plants to produce electricity from thousands of small rivers and streams that are often in the wilder corners of the National Forests. As encouraged by the Public Utilities Regulatory Power Act (PURPA), the Federal Energy Regulatory Commission (FERC) can issue permits to private individuals for such projects. Applications threaten dozens of roadless areas, mostly on the West Coast.

The best tool for protecting free-flowing rivers and streams is designation as part of the National Wild and Scenic Rivers System or a state river protection system. The national system was established by the 1968 National Wild and Scenic Rivers Act; many state systems have since been established as well. Although Wild and Scenic River designation has been inadequately utilized during the past twenty years, conservationists are gearing up a major new national campaign. Inclusion in the system generally places only a quarter-mile-wide zone on each side of the river under protection, but it does safeguard a river from dams and other development that would modify its free-flowing character.

Power-line and pipeline corridors. Associated with the extraction of energy sources is the construction of pipelines, power lines, coal-fired power plants, and so forth. Power lines and pipelines slice across the backcountry and divide many units of the Big Outside from one another. Irrigation canals, aqueducts, and power lines from hydropower and water-storage dams cut across many remote sections of the country, dividing large roadless areas from one another. More lines are projected, and new transmission corridors will be proposed through large roadless areas.

Slob hunting. I am a hunter, and offer no apology for it. Hunters have been among the most effective wilderness and wildlife conservationists. This does not negate, however, the impact of the slob hunter (and of poor public policy catering to slob hunters) on wildlife and wildlands. The popular conception of the hunter as a fat, drunken bumpkin or urban good ol' boy cruising the backwoods in a jeep, armed with little natural history or appreciation of nature but plenty of ammunition, is all too true. Slob hunters fall into several categories.

The market hunter. A booming black market exists for body parts of Black Bears (gall bladders, paws), Elk antlers and teeth, Grizzly claws and skins, and so on, sought by practitioners of oriental medicine, collectors, and sexually deficient oddballs who believe these items possess aphrodisiacal or restorative properties. Big bucks can be made both by individuals and well-organized rings. Overworked game wardens catch only a handful of these dangerous criminals.

Apologists for hunting claim that no species has become extinct because of hunting. In reality, market hunting and "game hogging" for American Alligator, Bison, Gray Wolf, Elk, Bighorn, Passenger Pigeon, Carolina Parakeet, Wild Turkey, and numerous species of waterfowl and shorebirds played as major a role as did habitat destruction in extirpating or drastically reducing these species.

The road hunter. This is the stereotypical hunter. He wants to drive his jeep, trail bike, or ATV to where he'll shoot his freezer meat or anything else that moves. He opposes designation of Wilderness Areas because he can't drive in them. He doesn't like predators because they're eating *his* deer, Elk, or Moose. The Arizona Wildlife Federation, for example, generally opposes Wilderness designations because it largely represents this type of hunter. On the other hand, the Idaho Wildlife Federation supports more Wilderness than does the Sierra Club because it's made up of *real* hunters—men and women who know that wilderness provides hunting at its best.

The "gut hunter." These fellows shoot at any game they see, regardless of the distance. Firing countless rounds at an Elk or deer several hundred yards away, gut hunters miss more often than not. Too often, however, they succeed in gut-shooting a critter that then wanders off to die in agony.

The poacher. These people also need roads. They shoot without respect, and outside the law.

The trophy hunter. Some trophy hunters are conservationists, and support protection of the land. Others, such as many in the Foundation for North American Wild Sheep, want to eliminate predators and have road access everywhere. Trophy Bighorn Sheep hunters are usually wealthy, and are leading opponents of Wilderness designation for areas in the California Desert and in Arizona National Wildlife Refuges. Other trophy hunters concentrate on Mountain Lion, Grizzly, and other top-level predators.

The trapper. Trapping is legal, and encouraged by fish-and-game departments, in most states. Not only is it cruel, but it is usually done from road vehicles or ATVs. Trapping targets Bobcat, Lynx, Marten, Mink, Fisher, River Otter, and other predators with low reproductive rates. Trapping upsets the normal predator-prey balance. Trapping caused the near-extermination of Beaver from much of the United States, and trapping today continues to keep Beaver populations at an unnatural low. (Trapping by Native Americans in Canada and Alaska is arguably another matter.)

The "put-and-take" fisherman. While fly-fishing for native, naturally reproducing fish is one of life's higher callings, many fishermen just want to catch their limit (or exceed it, if no game warden is about). They are a powerful lobby that has created a fish-farming orientation among state wildlife agencies. Non-native, hatchery-reared fish that compete with natives have been introduced throughout the United States. Put-and-take fishermen have caused the introduction of trout to many high-country lakes and tarns in Wilderness Areas that did not naturally contain fish. This has upset delicate aquatic ecosystems. Lake and riverine fauna have been more transformed than any other in the United States. Put-and-take fishermen have been as much to blame for this as have polluters and dam builders.

Slob hunters of all flavors oppose Wilderness designations, create roads, kill excessive numbers of wildlife, and help turn the back-country into a game farm.

Wildlife "management." The U.S. Fish and Wildlife Service and state game-and-fish departments are partially composed of outstanding professionals who love wildlife and wilderness. They are disciples of Aldo Leopold, who founded the science of wildlife management and argued for the "land ethic." Unfortunately, many wildlife agencies are controlled by political appointees who represent slob hunters or welfare ranchers, and are staffed by arrogant bureaucrats who believe in running game farms on the public lands for their constituency of road hunters and put-and-take fishermen. This kind of wildlife manager supports clearcutting, vegetative manipulation, predator control, and roads, because these often favor hunters' favorite species, such as deer, or provide hunter access. This kind of wildlife manager stocks lakes and rivers with exotic fish or hatchery-reared fish because such stocking sells licenses and brings more money to the department. This kind of manager promotes hunting of top-level carnivores such as Mountain Lion and Grizzly because politically powerful ranchers and trophy hunters demand it. This kind of wildlife manager releases non-native birds like pheasant and Chukar because quail and grouse don't provide enough hunting. In bizarre cases, such as occurred with the New Mexico Game and Fish Department in the early 1970s, exotic species such as Oryx, Barbary Sheep, and Iranian Ibex have been released on the public lands to create huntable populations for which high license fees are charged.

Of course, we must understand that any bureaucracy promotes programs that create work for itself. Not until wildlife managers are forced to realize that their job is not to maximize the production of deer, pheasant, trout, or other "desirable" game species, but to maintain wildness and native diversity, will the profession live up to the standards Aldo Leopold established for it.

Eradication of species. With rare exceptions, every ecosystem in temperate North America has lost key species. In the East, Cougar, Gray Wolf, and Elk have virtually disappeared. In the heartland, Bison, once 60 million strong, are gone. In the West, Grizzly and Gray Wolf have been largely extirpated. Along the northern border, Wolverine, Woodland Caribou, Lynx, and Fisher are ghosts, lingering only in the wildest places. In the Southwest, the tropical cats (Jaguar, Ocelot, Jaguarundi) are shadows seldom seen. Bighorn Sheep, Black Bear, and Wild Turkey have been severely reduced in numbers wherever they once ranged. Riparian systems have had their native fish and invertebrate faunas so altered that exotics now dominate. Without the sensitive, wilderness-dependent species, wilderness is a hollow shell. Without the top carnivores, the dynamic balance no longer exists. What will become of the deer, without the wolf to whittle its swift legs? Is the mountain still alive without the bear?

Extirpation of native species is perhaps the most insidious tool of wilderness destruction. For conservationists, it is not enough merely to protect the land from the bulldozer and chain saw. We must return the rightful inhabitants to their homes.

Introduction of exotics. As native species have disappeared, as the balance has been upset, exotic, weedy species have invaded, thereby changing whole ecosystems. Fragmented ecosystems, with smaller cores and greater area in "edge" conditions, are highly vulnerable to invasion by such species. Many of these exotics were deliberately introduced by unthinking people. Most of the grasses in California are exotics. The salt cedar (tamarisk), from the Middle East, crowds out cottonwood and willow in the Colorado and Rio Grande drainages. House Sparrows, Rock Doves, Starlings, and Chukars have taken over the air and fields in many places. Spotted Knapweed chokes out native grasses in the Northern Rockies of Idaho and Montana. Alfred Crosby, in his brilliant and ground-breaking book *Ecological Imperialism,* argues that we have created "neo-Europes" in temperate areas around the world. The deliberate introduction of Crested Wheatgrass by the BLM in the Great Basin is probably the major current attack on the Big Outside from this angle.

Suppression of wildfire. Naturally occurring wildfire (generally started by lightning) is an important component of most ecosystems in the lower forty-eight states. Periodic fire is necessary to cause certain seeds to sprout, recycle nutrients, maintain prairies, thin out vegetation, and accomplish other ecosystem services. The suppression of wildfire (the "Smokey the Bear Syndrome") has degraded wildernesses throughout the country. The fires that raged through the Pacific Coast forests in 1987 and across Yellowstone in 1988 were simply inevitable natural events that accomplished much ecological good. The Forest Service and Park Service have begun to acknowledge the valuable role of fire in wilderness ecosystems and have, in some cases, established "let burn" policies for natural fire in Wilderness Areas. Unfortunately, when commercial timberlands or private property outside Wilderness Areas are threatened, full-scale fire control, including bulldozers and slurry bombers, is unleashed. Fighting a forest fire or grass fire is nearly always more destructive than letting it burn because the use of heavy equipment like bulldozers and the cutting of fire lines causes greater erosion and disruption of ecosystemss.

Off-road vehicles (ORVs). Twenty-five years ago the problem of ORVs scarcely existed. Jeeps, four-wheel-drive pickups, dirt bikes, and snowmobiles were rare. Motorized tricycles and other all-terrain vehicles (ATVs) had not yet been invented. Today, however, millions of these infernal machines are piloted by boys trying to exorcise the demons of their puberty, or by people who want to "get into the backcountry" to hunt, fish, trap, poach, treasure hunt, prospect, or camp, but are not willing (or in good enough shape) to hike in. ORVs destroy vegetation, disrupt wildlife, erode the land, foul streams and air, and provide access to pristine areas for people who do not respect such places.

The disturbing question is why land managers allow ORVs at all. Both the BLM and the Forest Service have full power to restrict or prohibit off-road travel. Presidents Nixon and Carter both issued Executive Orders giving federal agencies explicit authority to control ORVs. The vast majority of the over 300 million acres of

National Forest and BLM land in the lower forty-eight states, however, is open to ORVs—not just on jeep routes or dirt-bike trails, but *cross-country*. ORVs carve thousands of miles of new, low-standard roads into roadless areas of the public lands every year. At the very least, vehicles should be restricted to designated roads, with all cross-country travel absolutely banned.

Why is this not done? Two reasons come to mind. First, many Forest Service and BLM employees and managers use ORVs in the backcountry themselves, and therefore identify with other recreational ORVers. Second, ORVers are well organized and vocal. They scream like scalded hogs when they are restricted in any way from exercising their "constitutional rights" to drive wherever they wish. Although the public dislikes ORVs and their use on the public lands, this rude minority gets its way.

Four-wheelers, dirt bikers, and other motorized recreationists present the strongest opposition to protection of the California Desert. They represent a large anti-Wilderness constituency in other areas as well. Snowmobilers are a similar stumbling block to protection of wildlands in the northern states and Rockies.

Industrial tourism. Outdoor recreation has become a big business. Large corporations, land developers, and small businessmen operating in National Parks (concessionaires) and "gateway" towns (including local chambers of commerce) have exploitative attitudes toward wildlands that rival those of loggers or miners. National Park administrators rank their "success" by the number of visitors they host (as indicated by Yellowstone National Park's declaration of its plans to advertise heavily to get visitation up again after the adverse publicity of the 1988 fires). A large number of outdoor recreationists loathe "roughing it," and demand full hookups (electricity, water, sewage) for their travel trailers or motor homes, also known as recreation vehicles (RVs).

RV campgrounds, condominiums, second-home subdivisions, resorts, golf courses, ski areas, tennis clubs, recreational reservoirs, marinas, scenic highways, visitor centers, motels, and access roads serve these industrial tourists. In doing so, they usurp prime winter habitat for Elk and Bighorn, indirectly cause the deaths of Grizzly Bears in Yellowstone and Glacier, create air pollution and traffic jams in remote areas, replace native vegetation with exotics, destroy wild rivers and streams, overfish and overhunt (thereby encouraging the game farming mentality), and bring far too many inexperienced people into delicate ecosystems.

Large roadless areas are threatened by ski-area development in California and Colorado; ambitious wilderness-recovery plans are being torpedoed by condos and recreation subdivisions in New England; the survival of the Grizzly in Yellowstone is jeopardized by RV campgrounds; and water skiers zip over the drowned Glen Canyon. In every section of the country, wilderness and wildlife are trampled underfoot by various manifestations of industrial tourism.

The National Park Service has many fine employees (as do the Forest Service and BLM) who value the wild and answer a calling to protect it. Unfortunately, some of the top administrators have lost touch with the wild nature their Parks were established to preserve, and have become, in many cases, leading threats to the Parks. Developments such as Fishing Bridge and Grant Village, along with arrogant mismanagement of Grizzlies, have disrupted the ecological integrity of Yellowstone National Park. The tacky urban center of Yosemite Valley is a national disgrace. Commercial outfitters dictate policies on river running in the Grand Canyon and other Parks, and lock out private boaters. Corporations offering "scenic overflights" are given free access to skies over Parks by Park Superintendents who enjoy buzzing around in helicopters, too. Concentrating on scenic views and visitor services, Park Superintendents have allowed development in sensitive ecosystems. The primary constituency of the Parks is not the residents—wildlife—but local chambers of commerce, concessionaires, and the motorized tourist. Indeed, concessionaires (often subsidiaries of multinational corporations) have largely usurped management of popular Parks from the Park Service, and run them to maximize their profits.

Unless the National Park Service can get back on track with a philosophy of ecosystem management, and kick out the concessionaires, the National Park ideal which the United States gave the world will become a cruel hoax.

Wilderness recreation. One would think that those who take the time to hike, float, or horsepack into Wilderness Areas would seek to protect the pristine quality of the land. Most do, but a minority, often locals on horseback but sometimes urban backpackers, show no respect for the Wilderness. Fire rings without number, semi-burned aluminum foil, toilet-paper "flowers," hacked green trees, empty Coors beer cans, discarded fishing line, soap in streams and lakes, horse tethering in campsites or hobbling around lakes—all are the calling cards of wilderness slobs. In extreme cases, commercial hunting guides and packers establish semipermanent Wilderness camps that resemble small towns. Some outfitters have even packed in prostitutes to service hunters in places such as Wyoming's Teton Wilderness Area.

Wilderness recreationists who fail to practice sensitive backcountry ethics should be fined and banned for specific periods from entering Wilderness Areas. Commercial outfitters should be carefully supervised, and should have their permits yanked for trashing Wildernesses. The Forest Service, Bureau of Land Management, Park Service, and FWS need to hire more (and better qualified!) Wilderness rangers to enforce proper backcountry use.

Individually and in concert, these methods of wilderness destruction fragment the remaining wild places. By chopping large ecosystems into smaller pieces, not only do these pieces become extremely vulnerable to disruption, but they can no longer support the full array of native animals and vegetation that they once supported as larger areas. Understanding the factors that cause the destruction of wilderness is the first step to preventing such destruction. None of the remaining roadless areas in the United States is large enough to stand alone. None is large enough to maintain the minimum viable populations of wide-ranging predators.

Identifying the remnants of the Big Outside is the next step.

History shows us that even the most sublime wild place cannot be protected from destruction, no matter how egregious or lunkheaded, unless defenders of the wild know about it.

In 1956, conservationists accepted a compromise on the Colorado River Storage Act, which canceled a huge dam on the Green and Yampa Rivers in Dinosaur National Monument in favor of one on the Colorado River at Glen Canyon. Except for a few pioneer river runners like Ken Sleight and Katie Lee, no one objected. The conservationists who made that compromise knew the canyons of Dinosaur, but they didn't know Glen Canyon. David Brower has said that that compromise was the greatest mistake he ever made. It was the tragedy of "the place no one knew." The damming of Glen Canyon cut the heart out of the largest roadless area in the United States.

Other great roadless areas have similarly been destroyed because they were unknown. The southern Nevada desert, described by Bob Marshall as the finest desert wilderness he ever visited, also was neglected. It became an atomic bomb testing range. In too many other cases, conservationists have not fought for areas, large and small, because they were known merely as blank spots on maps.

May we never again lose the place no one knew!

9 | My Heroes Used to Be Cowboys

Oh give me a home, where the buffalo roam, where the deer and the antelope play. . . .

—traditional

I grew up wanting to be a cowboy. My great-grandparents on my mother's side homesteaded in New Mexico before statehood. I remember the stories my grandmother used to tell me about the hoe that was always by the door of the dugout—whenever you went out, you took it to kill rattlesnakes. And about her great uncle, Marshal Billy Crawford, who left Missouri in the 1880s for New Mexico and was shot in the back while trying to arrest Billy the Kid. In reality he was a city detective in Albuquerque and was shot in a sleazy hotel in 1888, while investigating the theft of some Navajo sheep. I discovered that while going through microfilms of old newspapers at the library, but my grandmother never believed me.

My daddy was from Texas, and I remember his story of General Patton chewing him out at Orly Field, Paris, during World War II. Pop was striding across the runway from his plane, resplendent in scarf, helmet, leather jacket, goggles, and . . . cowboy boots. Patton thought he was out of uniform, but Pop had a forged medical document saying he could wear only cowboy boots because of a foot problem.

As for me, I was born in Albuquerque, back in the days when it was still a Western *town*. When I was taken back to Kentucky to be presented to some of my father's family when I was a year and a half old, it was the cowboy boots and hat I wore that impressed them. In the sixties, while my friends at the University of New Mexico listened to the Beatles and the Rolling Stones, I tuned in to Johnny Cash, Ernest Tubb, Merle Haggard, and Loretta Lynn.

Maybe it was all those hours watching Hopalong Cassidy on the silver screen, but, damn it, I wanted to be a cowboy.

After leaving college with a liberal-arts degree, I needed a job, so I became a horseshoer and was elected president of my farrier class at New Mexico State. In the early seventies, my wife (who represented the same redneck stock from which I issued) and I started a mulepacking business in the Gila Wilderness to supplement my meager income as The Wilderness Society's Southwest field representative. Our dream, though, was to be cowboys. We asked my wife's father, who was a county agent for the Farmer's Home Administration in New Mexico, about loans to buy a ranch. He laughed and said that there was no way a young couple could make it, that we would never be able to pay off a ranch. Ranching in the West was on the way out, he said. If you didn't inherit a big spread, it only made sense if you were some rich sonofabitch or a corporation looking for a tax write-off.

Well, maybe I couldn't be a rancher, but I still liked those broken-down old farts, with their incredibly sweat-stained hats, leathery, beat-up faces, big, rough hands, stove-up bowlegs, and scuffed boots. After all, because they were close to the land, they ought to be conservationists. Through the seventies I worked at building bridges to the cowboys. Seventy-five-year-old Jewel Smith told me I was the only man in Catron County who could keep up with her bucking hay. I branded cows with them, I sheared sheep with them, I commiserated with them about the bureaucrats with whom they had to deal, I tried to tell them how Wilderness designations wouldn't hurt them, and I played to their already overinflated egos by telling them they represented the spirit of the frontier. The compromises we worked out were always one-sided—they won. The most I ever got was an overcooked beef dinner and a sore back.

Finally, even a dumb horseshoer who's been kicked too many times in the head starts to look around and ask questions.

And I did.

And what I found was that cattle and sheep graze on most public lands in the West. The effects of ranching on the Western landscape are pervasive, shaping the look of the West and causing more ecological damage than any other single agent. Field ecologist and author George Wuerthner says, "Domestic livestock grazing has been the greatest environmental calamity ever to befall the western United States." Some authorities believe that livestock grazing has reduced the biological productivity of the West to roughly one-half its pre-grazing level. Andy Kerr of the Oregon Natural Resources Council likens grazing to clearcutting the federal range on an annual basis.[1] Let's look at the ways livestock grazing harms wilderness.

Elimination of predators. Before the coming of Europeans, several subspecies of the Gray Wolf *(Canis lupus)* inhabited most of the West. Between 1850 and 1900, two million wolves were poisoned, shot, or trapped to make the West and the Great Plains safe for livestock. By 1880, the Gray Wolf was gone from the Great Plains. The early decades of the twentieth century saw the extermination of the Texas Gray Wolf *(Canis lupus monstrabilis)* and the New Mexican Wolf *(Canis lupus mogollonensis)* in 1920, the Great Plains Lobo Wolf *(Canis lupus nubilus)* in 1926, the Southern Rocky

1. A good summary of the impacts of livestock grazing is Denzel and Nancy Ferguson's book *Sacred Cows at the Public Trough* (Bend, Ore.: Maverick Publications, 1983). The best quick overview of the impacts of livestock grazing on the public lands is George Wuerthner's article "Counting the Real Costs of Public Lands Grazing," in *Earth First! Journal* (August 1989). Many of the figures cited here are taken from Wuerthner's excellent article and from the Fergusons' groundbreaking book. For a detailed study of public-lands ranching and related issues, see Lynn Jacobs's *The Waste of the West*, available from Lynn Jacobs, P.O. Box 5784, Tucson, AZ 85703.

Mountains Wolf *(Canis lupus youngi)* in 1940, and the Cascade Mountains Brown Wolf *(Canis lupus fuscus)* in 1950.[2] The Mexican Wolf *(Canis lupus baileyi)*, the most distinct subspecies of Gray Wolf in North America, is fast approaching extinction. Although effectively exterminated in Arizona and New Mexico, this Lobo prospered in the Sierra Madre of Mexico until recent decades, when cattlemen's poison control programs—using 1080 (sodium mono-fluoroacetate)—began to take their toll. David E. Brown tells the grisly story of the cattlemen's campaign to rid the Southwest of wolves in *The Wolf in the Southwest: The Making of an Endangered Species.* [3] The rhetoric of the wolf-control program is chillingly similar to Hitler's "final solution" for the Jews—even to the extent of proposing a "wolf-proof fence" all along the Mexican border. Cattle and sheep ranchers are the most intransigent lobby today against restoration programs for the Mexican Wolf in the Southwest, and the Northern Rocky Mountains Wolf in Yellowstone National Park, Central Idaho, Northern Montana, and Washington State.

The Grizzly Bear, whose historic range included all of the West and the Great Plains except for the most arid regions in the Southwest and Great Basin, suffered an extermination campaign for the sole benefit of stockmen almost parallel to that suffered by the Gray Wolf, except that Griz populations hung on in Yellowstone National Park and northwestern Montana. Although eradicated from Arizona and New Mexico by the mid-1930s, the Mexican Grizzly persisted in tiny numbers in the remote backcountry of Mexico's northern Sierra Madre until 1960. Then ranchers launched a final assault, poisoning, trapping, and hunting the Mexican Grizzly to extinction. As with the wolf, ranchers are the primary lobby against proposals to reintroduce Grizzlies in large, remote Wilderness Areas in the West. A leading cause of death for the critically small Grizzly population in Yellowstone is shooting by sheep ranchers in the adjacent Targhee National Forest of Idaho.

2. Dates of extinctions are taken from *Vanished Species,* by David Day (New York: Gallery Books, 1989 [revised]).

3. David E. Brown, ed., *The Wolf in the Southwest: The Making of an Endangered Species* (Tucson: University of Arizona Press, 1984).

The Mountain Lion would have been exterminated along with the Gray Wolf and the Grizzly, except that it is more difficult to trap or poison because it does not normally feed on carrion. Nevertheless, the number of Mountain Lions in the West has been drastically reduced by stockmen and government "control" programs for their benefit. An even larger cat, the Jaguar, once was native to the Southwest. It has been virtually extirpated by predator-control programs.

After destroying most large, wide-ranging predators in the West, ranchers and the government trappers and hunters working for them turned their attention to Bobcat, Black Bear, Golden Eagle, and, particularly, Coyote. In the late 1980s, ranchers began demanding that the federal government exterminate Common Ravens because they were, the ranchers claimed, preying upon calves and sheep!

The poison and leg-hold traps used to kill supposed livestock predators also kill many "non-target" species. In 1963, for example, the U.S. Fish and Wildlife Service reported killing 6,941 badgers, 24,273 foxes, 10,078 raccoons, and 19,052 skunks (none of which has ever been accused of killing cattle or sheep, so far as I know, even by hysterical and biologically ignorant ranchers). Not only do ranchers wish to eliminate any animal that might prey on cattle or sheep, but they also wish to destroy those who might eat grass and thereby compete with domestic livestock. The federal government's Animal Damage Control unit has poisoned millions of prairie dogs and other rodents for this reason. The probably imminent extinction of the Black-footed Ferret, which fed almost exclusively on prairie dogs, is due to such extermination campaigns for ranchers. Several of the last wild California Condors died from eating rodents killed by poisoned baits. (The secondary effects of certain "predicides" take an immense toll on carrion feeders.)

The Arizona Game and Fish Commission reports that predator-control programs and other impacts of livestock grazing are responsible for population declines of thirteen out of eighteen mammal and eleven of twenty-two bird species listed as Threatened or Endangered in that state.

The fears of stockmen concerning predation are vastly overblown and represent a basic fear of uncontrolled nature. Predators, more than any other type of wildlife, represent wildness. In their relentless hatred for predators, ranchers reveal their essential hatred for the untamed, the undomesticated. I remember once telling Jewel Smith about a friend seeing a young Black Bear treed by Coyotes in the Gila Wilderness. "Why didn't he shoot that li'l bear," she asked, "and the coyotes, too?" It was inconceivable to her that anyone would see a varmint of any sort and not kill it. My landlord for seven years in Glenwood, Ellsworth Tipton, was a fine old gentleman, then in his eighties, whose family had been among the first ranchers to arrive in the area after the Apaches were "pacified" in the early 1880s. He had tracked and attempted to kill every Mountain Lion that ever dared set foot on "his" ranch (virtually all National Forest land). His lifetime toll was more than twenty cats.

The full impact of this war of genocide against predators becomes clear only in the light of current ecological research. We now know that wide-ranging predators at the top of the food chain are the key indicator species for healthy ecosystems, and are often the most vulnerable members of any ecological community. Aldo Leopold wrote, after killing a wolf in Arizona:

> *I now suspect that just as a deer herd lives in mortal fear of its wolves, so does a mountain live in mortal fear of its deer. And perhaps with better cause, for while a buck pulled down by wolves can be replaced in two or three years, a range pulled down by too many deer may fail of replacement in as many decades.*

We retain these fearsome beasts as cultural icons—the University of New Mexico's football team, the Lobos, the Grizzly Bear on the state flag of California—but unless we share the land with them again, the West will become as tame spiritually and ecologically as Britain.

Competition with native herbivores. Commercial livestock grazing has caused a tragic reduction in numbers of Elk, Bighorn Sheep, Pronghorn, and Bison in the eleven Western states. Bison, originally a population of 5 to 10 million, are now down to a few thousand; Pronghorn, originally 10 to 15 million, are now 271,000; Bighorn Sheep, originally 1 to 2 million, are now 20,000; and Elk, once 2 million strong, now number 455,000.[4] While part of this decline has come from overhunting and habitat destruction by logging, mining, roads, and urban sprawl, most of it is due to competition with domestic livestock.

A key limiting factor for herbivore populations is winter range, those grazing areas at lower elevations, on south-facing slopes, and in river valleys, that generally remain snow-free in the winter. Most of the good winter range in the West is now private property and used for farming and hayfields for domestic livestock. But even on public lands, cattle and sheep are favored over wildlife. For example, in 1980, on the Lakeview (Oregon) District of the Bureau of Land Management, livestock were allotted 166,454 animal unit months[5] and wildlife only 10,916 AUMs. Ranchers in Nevada raised such an uproar over the reintroduction of Elk into the Jarbidge Mountains in the mid-1980s that the Nevada Department of Wildlife withdrew its proposal. The ranchers claimed the Elk would eat too much grass. In 1989, Mike Oden, a rancher on the Prescott National Forest in Arizona, was convicted of slaughtering one-third of the Elk herd in the area he leased for grazing. The Elk that wildlife agents found had been gut-shot with a small-caliber weapon so they would wander off and suffer "a lingering death." He feared the Elk were taking forage away from his cattle. He was fined a measly $6,507.

In the mid-1970s, ranchers in the Roswell, New Mexico, area mobilized against the Bureau of Land Management's plan to modify net wire sheep fences to allow for the passage of Pronghorn to reach seasonal water sources. The ranchers claimed, without support from

4. These figures do not include the Great Plains. There were an additional 50 million Bison, 25 million Pronghorn, and many Elk and Bighorn Sheep on the Great Plains. Decreases in numbers on the Plains are even greater than those given for the Intermountain West.

5. An animal unit month (AUM) is a standard measure of grazing pressure. It represents the grazing of one cow and her calf, or five sheep, for one month.

wildlife experts, that such modifications would allow Coyotes through the predator-proof fences. The Llano Estacado of southeastern New Mexico had formerly supported one of the healthiest Pronghorn populations in the world, but it crashed after the tightly woven fences were erected for sheep grazing on public lands in the area. I cut my eyeteeth on this ranching issue, and picked up cigar smoking to boot, when I came down to shear sheep for Bud Eppers, leader of the sheep-grower faction, in an attempt to build bridges between the livestock and environmental communities in New Mexico. As usual, the ranchers won on the Pronghorn fence issue.

Modification of native vegetation. The grazing of cattle and sheep has drastically altered natural vegetation communities and has led to the introduction of non-native grasses palatable only to domestic livestock. In the Southwest, livestock grazing combined with fire suppression has caused woody plants such as juniper, mesquite, and Creosote Bush to invade former grasslands. In the Great Basin, the same factors have allowed Big Sagebrush to take over grasslands. Cheat Grass, an Old World species that first appeared in Pennsylvania in 1861, spread on its own to the Northern Rockies and became the dominant plant on millions of acres of rangeland during the twentieth century, thanks to the destruction of native grasses by livestock. In recent decades the Bureau of Land Management has planted Crested Wheatgrass, an exotic, throughout the same region for cattle forage, and ripped up native vegetation by the roots with hellish plows dragged by bulldozers. Neither of the foreign grasses meets the needs of the wildlife species, ranging from Elk to songbirds, that depend on grasslands. In 1976 it was estimated that 75 percent of the Western range was producing forage at less than half of its potential, owing to the effects of domestic livestock grazing.

Erosion and watershed destruction. The destruction of native vegetation by livestock has led to sheet erosion and gully erosion, sweeping away most of the topsoil of the West. In arid and semiarid lands, soils are often bound together and capped by a cryptogamic soil-lichen association. Trampling by cattle destroys this delicate layer and opens the soil to massive erosion from rainfall. A few examples cited by Denzel and Nancy Ferguson in *Sacred Cows at the Public Trough* (see pages 64–70; see 89*n*) should suffice to give some indication of the overall impact of livestock grazing on soils:

■ Only 5 percent of the total runoff in the lower forty-eight states comes from the Western public lands, but they produce 320,-000 acre feet of silt annually—more than the combined sediment loads carried by the Mississippi and Colorado Rivers.

■ Five hundred million tons of soil erode annually from Western public rangelands—many times faster than the soil is replaced.

■ The Rio Puerco, west of Albuquerque, was once described as the "bread basket" of New Mexico, and supported thriving villages and farms. Owing to overgrazing, the watershed lost 1.1 billion to 1.5 billion tons of soil between 1884 and 1962. The area is virtually uninhabited today, and ranchers in the half-million-acre area netted only $9,527 in total profits in 1974. In the mid-1970s the U.S. Army Corps of Engineers proposed a $120-million silt-retention dam downstream to keep sediment from entering the Rio Grande. It was never built.

■ Great concern is expressed about anthropogenic desertification in the "backward" nations of Africa, Asia, and Latin America, but some of the worst desertification in the world is occurring on land managed by the United States Forest Service and Bureau of Land Management; 225 million acres of land in the lower forty-eight are undergoing severe or very severe desertification.

■ In a 1985 inventory of 118 million acres, the BLM found 71 percent to be in an unsatisfactory range condition.

Damage to streams and riparian areas. Commercial livestock grazing has severely degraded streams and riparian vegetation throughout much of the West. Using the government's own figures, the Fergusons point out the evidence:

■ The BLM in Nevada says, "Stream riparian habitat where livestock grazing is occurring has been grazed out of existence or is in a severely deteriorated condition. Within the state 883 miles of streams [are] either deteriorated or declining."

■ Utah: Streams in grazed areas are 173 percent wider than on ungrazed sections of the same streams—indicating severe erosion.

■ Montana: Soils along an ungrazed portion of a stream retained 772 percent more moisture than did soils in a grazed portion.

■ Montana: Streamside vegetation was 76.4 percent greater along an ungrazed portion of Rock Creek compared to a grazed portion. The ungrazed section produced, by weight, 336 percent more trout.

■ Utah: Within five years of fencing out cattle along Big Creek, vegetation recovered and trout increased 570 percent. Trespass grazing by livestock for six weeks essentially returned conditions to what they had been before cattle were excluded.

In the dry Southwest and Great Basin, cattle tend to concentrate along streams because of water availability. Riparian areas are thus more heavily grazed than any other. Moreover, certain riparian plants are "ice cream species" particularly favored by cattle. These include young willows, cottonwoods, and sycamores. In many riparian areas of the Southwest, there is little reproduction of such trees, due to cattle. On a seven-day backpack trip in the rugged Blue Range Primitive Area in eastern Arizona, I came upon a fence crossing the Blue River. Upstream, where cattle grazed, there were no tree seedlings at all on the gravel bars, mudflats, and terraces along the river. Downstream, where cattle had been removed for several years, young cottonwoods, willows, and sycamores were lush.

Researchers have found that 75 to 80 percent of all vertebrate species in Arizona and New Mexico depend on riparian areas for at least part of the year. Less than 3 percent of the original riparian community remains in Arizona; the willow-cottonwood riparian community is the rarest of 104 major plant types in the United States. This ecosystem is home to one hundred state and federally listed Endangered and Threatened species in Arizona and New Mexico.

Destruction of streams has had a devastating impact on native fish species. Of thirty-two species native to Arizona, five are extinct and twenty-one are Endangered, Threatened, or being considered for listing.

Human and wildlife health. Diseases from domestic sheep have been a major factor in exterminating wild Bighorn Sheep from much of the West. One of the largest herds of Rocky Mountain Bighorns in the United States was in the Eagle Cap Wilderness Area of Oregon. Diseases borne by domestic woollies reduced the herd by 75 percent. A reintroduced herd of Bighorns was completely wiped out by an epidemic from domestics in 1988 in California's South Warner Wilderness Area. The widespread introduction of the giardia parasite in wildland streams and lakes in the West has been largely caused by domestic cattle. Where once a hiker or camper felt secure drinking from a sparkling wilderness stream, now a debilitating parasite lurks. Welfare ranchers commonly blame "hippies" for spreading the disease to their cows.

Diminishing recreational values. Besides the health hazard of giardia, grazing by cattle and sheep spoils the aesthetic experience of many outdoor recreationists, ranging from roadside picnickers to deep wilderness trekkers. In many backcountry areas, it is almost impossible to find a campsite not fouled by cow pies. Sport fishers have become strong opponents of livestock grazing on public lands after recognizing the devastating impact of grazing on trout. Many outdoor recreationists find that livestock grazing decreases "watchable wildlife." In much of the West, vegetation is a key part of the scenery recreationists come to enjoy—Saguaros and other cacti in the Southwest, cottonwoods along rivers, flowing grasslands. Livestock have severely damaged those visual resources as well.

Roads and other developments in wildlands. In non-timbered areas, most developments on public lands—roads, fences, juniper chainings, windmills, pipelines, stock tanks, etc.—are for the sole benefit of a few welfare ranchers. By ending public-lands ranching, tens of thousands of miles of dirt road could be closed on BLM and National Forest land and millions of acres of public land allowed to return to wilderness conditions.

Compromise of National Parks, Wildlife Refuges, and Wilderness Areas, and restrictions on access to other public lands.
Many Wilderness Areas, National Parks, National Monuments, and National Wildlife Refuges are grazed by cattle or sheep. The Wilderness Act specifically permits the continuation of livestock grazing where established prior to designation of the area as Wilderness. Congressional language attached to the 1980 Colorado Wilderness bill extended the rights of commercial graziers in Wilderness Areas, permitting them even to use motorized vehicles and equipment to manage their herds. Capitol Reef, Grand Teton, and Great Basin National Parks, Dinosaur and Black Canyon of the Gunnison National Monuments, and Glen Canyon National Recreation Area are legally grazed. Only the death of the grazing permittee ended abusive cattle grazing in Organ Pipe Cactus National Monument in the late 1970s. This unit of the National Park System preserves the most diverse part of the Sonoran Desert in the United States and receives as little as three inches of precipitation a year. Trespass grazing by cattle is a common problem in National Parks and Monuments that otherwise prohibit it.

The sacred cow rules even National Wildlife Refuges. Wuerthner writes, "Grazing also occurs on most of the large National Wildlife Refuges in the West. Of the 109 Wildlife Refuges in Fish and Wildlife Service Region Six—Montana, Wyoming, Colorado, Utah, Kansas, Nebraska, North and South Dakota—103 are grazed."

Two hundred fifty million acres of National Forest and BLM land in the West are grazed: 89 percent of the BLM's total holdings and 69 percent of the Forest Service's. The lands not grazed are heavily timbered, rock and ice, too arid, or otherwise impossible to graze.

The West's 36 million acres of state-owned "ranchland" are even more overgrazed by livestock and dominated by stockmen than are

BLM ("Bureau of Livestock and Mining") lands. In several Western states, grazing permittees can legally lock out all other would-be users, including hikers, hunters, fishers, bird-watchers, campers, and picnickers. For the 10 million acres of state-owned land in Arizona, for example, public access is restricted to ranching leasees, and to holders of valid hunting or fishing licenses, and then only during season—the Arizona Game and Fish Department has to pay the State Land Office for that privilege!

Ranchers are so used to treating federal grazing permit lands as their private property that they commonly try to restrict access to BLM and National Forest areas by putting up No Trespassing signs and illegally locking gates. In many cases, road access to an area of National Forest or BLM land goes right through a rancher's front yard, and he does his best to intimidate the public into not using that access.

And, throughout the West, public-lands ranchers are the most vocal and militant lobby against environmental protection and Wilderness designation.

Perhaps the most telling comment about the effects of ranching in the West comes from a range expert in New Mexico who in 1985 flew over the Trinity Site, where the first atomic bomb was detonated little more than forty years earlier. He said that the range condition twenty yards from ground zero was better than over 90 percent of New Mexico.

So much for the myth of the rancher as conservationist.

But what of economics—and food production? I'll admit that I'd just as soon sink my teeth into a bloody slab of beef as into anything but wild meat. I'm no bleeding-heart vegetarian. However, public-lands ranching does not significantly contribute to food production in this country. Only 3 percent of America's total beef production comes from *all* U.S. public land—federal, state, and local. Three percent. That's less than the normal yearly fluctuation in production. Nevada, a cow state? Hell, Arkansas runs twice as much beef as Nevada! Studies by wildlife biologists indicate that more pounds of red meat could be produced on a sustained-yield basis from

Western public lands through increased wildlife production if the cows and sheep were removed. I'd rather get my steaks from an Elk back in the wilderness than wrapped in plastic at Safeway, any day of the week.

Economically, the Forest Service and the Bureau of Land Management spend several times more supporting the livestock industry than they receive in grazing fees. In Oregon, in 1985, the BLM spent $4,777,653 on range management and received $1,298,783 in grazing fees. Those figures are typical. Although it is difficult to peg all the costs of public-lands grazing, experts estimate that the Forest Service and the BLM lose over $100 million a year on their grazing programs. When erosion, lowered recreational values, loss of bio-diversity, and numerous other hidden costs are factored in, the subsidy to the livestock industry grows to gargantuan propor-tions—very roughly, $2 billion annually, or $66,666 per public-lands rancher. The proud, independent public-lands rancher as the paragon of the free-enterprise system? Forget it; he's a welfare bum. I heard one good ol' boy state at a grazing-fee hearing a few years ago, "Open bidding would destroy the very concept of free enter-prise."

So, commercial livestock grazing on the public's land can't be justified environmentally, economically, or agriculturally. Why is it there?

Politics.

The cattle and sheep interests have, for a century, been the most potent lobby in the Western states. Some state legislatures are more filled with pointy-toed boots than with law degrees. This despite the fact that only 30,000 individuals and businesses hold grazing permits for Western public land! For example, of the 4 million residents of Arizona, only 3,792 raise livestock, and only 1,323 of those hold grazing leases for federal lands. Many of these 30,000 permit holders are only part-time ranchers, and they run livestock on public land for an average of only four months each year. They hold regular jobs in town, or own a gas station, and play cowboy on the weekend. In other cases, their wives work and bring home the real paycheck. Nonetheless, ranchers are the landed gentry of the West, our self-proclaimed nobility, and they expect to be treated as such. (An amazingly high percentage are wealthy, and many are millionaires.)

Although the stockmen's control is slipping as opposition grows, efforts to reform ranching on public lands usually fail. When the Department of the Interior, during Jimmy Carter's administration, began a few tentative steps toward reform, the protest from the cowboys was so loud you could hear Interior Secretary Cecil Andrus's knee joints pop as he tried to do a quick about-face.

Cows and sheep don't belong on public land, period. But how do we get rid of them? One lobbyist for the welfare ranchers says to be patient; the range livestock industry will be dead in twenty years. I don't want to wait that long to see the restoration of the West. I want to see wolves and Elk and streams and grasslands coming back in my lifetime.

Another difficult matter to consider is that in spite of the many obnoxious, overbearing cattle and sheep barons, pushing their well-marbled guts and big belt buckles around the halls of Interior and the state legislatures, there are some damn fine folks trying to force a living from ranching just like their parents and grandparents before them (although less than 15 percent of permittees retain the original permit). Despite all I've said, there's a good reason for my dreaming of being a cowboy—it's a fine life, lived by a few fine people. Unfortunately, on public land in the arid West, it just doesn't make sense—for the land, or for the taxpayer's pocketbook. I have few problems with small producers on permanent pastures on private lands in the East or West. This is where beef should be produced.

It's time conservationists expanded the terms of reference for the debate over public-lands grazing. I propose four possible routes for the elimination of cows and sheep from the public land.

Establish open bidding for grazing rights. Currently, a grazing permit on Forest Service or BLM land is tied to a private property *base.* The federal lands grazing permit, in effect, becomes part of a ranch that may also include private and state grazing lands. When the private base property is sold, the federal permit goes along as part of the ranch. Ranchers even borrow money on the basis of their federal permit.

By federal law, grazing fees for federal permittees are kept extremely low. In 1990 the grazing fee charged by the BLM and Forest Service was $1.81 per AUM. In other words, for a dollar eighty-one, a rancher can leave a cow and her calf on federal land for a month. By comparison, the average AUM rate for comparable private land from 1964 to 1984 was $6.87. Some military reservations offer grazing privileges on a bid basis. This grazing goes for several times what ranchers pay for BLM or Forest Service AUMs. A minor scandal erupted in the 1980s when it was discovered that about 10 percent of federal permittees were subleasing grazing rights to other ranchers for four or five times what they were paying the government, and pocketing the difference.

A bidding system, not tied to a base property, for all public land currently in grazing allotments would have several advantages:

■ The government would be paid closer to fair market value for forage and browse, eliminating at least part of the subsidy the taxpayer currently provides to welfare ranchers.

■ Because ranchers would no longer have what are essentially lifetime grazing privileges, and would not be able to treat federal lands as part of *their* ranches, federal agencies would have greater power to check overgrazing and other abuses. Other interest groups, including conservationists and recreationists, would have more influence as well.

■ Marginal or incompetent operators would be driven out of business, which is what a competitive marketplace is supposed to do.

■ Finally, conservation and wildlife groups could bid on federal grazing permits. You want free enterprise? Fine. For a couple of thousand dollars, a group of hikers, bird-watchers, fishers, hunters, or Deep Ecologists could get a grazing permit and elect to graze Elk or quail or trout—or wolves—or all of these. For no more than grazing permits cost, conservationists could block up million-acre chunks of public land and remove the cattle. This could lead to all sorts of interesting scenarios for wilderness restoration. For this to take place, a change in policy would be required to allow the permittee to take voluntary non-use of livestock grazing. Current regulations allow the government to take a permit away if the permittee does not graze livestock.

Establish an honest welfare system. One seemingly powerful argument for preservation of welfare ranching is community stability. Even though a cow county like that around Elko, Nevada, derives only about 5 percent of its total income from livestock grazing, myth has it that ranching is the backbone of the economy in many areas of the West. Fine. Let's continue the welfare. Let's just not subsidize the destruction of public land.

Determine the average *profit* derived from the public-land share of a rancher's operation over the last ten years, and send the rancher a check for that amount every year for the rest of his or her life. He could continue to run cows on the land he owns. But remove his cows from National Forest, BLM, National Wildlife Refuge, or National Park lands. Surprisingly, this would cost the federal government *less* than even the most direct traditional subsidies to welfare ranchers. It would also eliminate the hidden, but greater, subsidy—that of degraded rangeland, watersheds, wildlife populations, and wilderness.

Former grazing permittees could also be hired to remove unneeded ranching improvements such as cattle guards, windmills, fences, and pipelines; to close roads; to install erosion control structures; and to remove exotic plants and replant natives.

Buy 'em out. An alternative way of effecting the above would simply be to calculate the actual value of the man's public-lands grazing permit and buy it from him, say, by equal installments over the next ten to twenty years. An economist can figure out the relative advantages and disadvantages of the two methods. Heck, we could let a welfare rancher pick his own type of subsidy. It will still cost us and the land less.

Run buffalo instead of cows. In those areas, primarily on the Great Plains, where Bison were present, a less draconian alternative would be to require ranchers to run Bison on their federal permits instead of cattle or sheep. Bison are native, they eat less grass, require less water, aren't bothered by predators or severe weather, and their meat tastes better.

The cowboy, as we dream him, existed for only a short time—if at all. We kill what we love. Although there are exceptions, the typical modern rancher is far from the mythical cowboy of old. He wears similar clothes, still nursemaids cows, and may even have a horse—standing idly in the corral in favor of a pickup. There the similarity ends and reality intrudes. The cowboy was a serf. His liege lord was the big rancher. The Virginian. The cowboy was strangled on barbed wire, and trod underfoot by the rancher/nobleman. While there still are a few small, mom-and-pop operations running cows and sheep on the public lands of Nevada and New Mexico, too often federal grazing permits are owned by corporations, wealthy professionals looking for a tax write-off, and those descendants of ranchers who have built empires and swagger like English squires.

The cowboy was an ephemeral transient through the American wilderness, like the earlier mountain man. We cannot bring him back. But we can bring back the geography through which he rode—the Wilderness.

10 | Return of the Wolf

Relegating grizzlies to Alaska is about like relegating happiness to heaven; one may never get there.

—*Aldo Leopold*

In the Northern Rockies, the North Cascades, and the Southwest, conservationists have been cheered lately by the natural recolonization of formerly occupied range by once-extirpated predators. The formation of two active Gray Wolf packs in the Glacier National Park area of Montana, after a long absence of wolf activity, has been covered by the national media and has led to a field day of research opportunities for biologists. According to Grizzly Bear expert Doug Peacock, the sign in the area gives convincing evidence that a wolf pack has a far greater ecological presence than do individual wolves.

Increased Grizzly Bear sightings in Washington's North Cascades have similarly exercised the media, biologists, and conservationists. Where once it was believed that only a handful of Grizzlies drifting across the border were using the North Cascades, it now appears that there may be a growing population of at least thirty big bears resident in the area.

Far to the south, the evidence is that increasing numbers of Ocelots, Jaguars, and Lobo Wolves are using the Southwestern borderlands from the lower Rio Grande Valley and Big Bend in Texas to the Bootheel of New Mexico to the Sky Islands and mesquite bosques of southeastern Arizona.

I am as excited and pleased as anyone that the wolves, Grizzlies, Jaguars, and Ocelots are returning, but I perceive a dark lining to this shining cloud.

For many years, the wilds to the north in British Columbia and Alberta, Canada, and to the south in Sonora, Chihuahua, Durango, and Coahuila, Mexico, afforded refugia for the great predators ruthlessly and efficiently exterminated in the States by cowmen, government trappers, and hairy-chested "sport" hunters. For forty years after the elimination of Lobo, Grizzly, Jaguar, and Ocelot from the Southwest, they persisted in the trackless deserts and mountains of northern Mexico. Occasionally a Lobo or spotted cat would drift across the border like a ghost, exciting the fears of the cattle barons and the imaginations of the rest of us.

Glacier National Park and the Bob Marshall country in northwestern Montana maintained the healthiest population of Grizzly Bears in the lower forty-eight states because it was not an island population like that of Yellowstone, but was constantly replenished by the seemingly limitless supply of silvertips to the north in Canada. The North Cascades in Washington held a small population of Griz, as did the border mountains of the Selkirks, Cabinets, Kettle Range, and Salmo-Priest for the same reason: As soon as the good ol' boys got one in their iron sights, another would slip in from the wild north across the line.

Even the fabled Timber Wolves of Isle Royale and the Boundary Waters in Michigan and Minnesota lasted until today because they, too, were directly connected to an unbroken tribe of wolves stretching to the Arctic.

Today, however, those wild nations to the north and south are no more. The guns, traps, poisons, cattle, chain saws, bulldozers— the tools of civilization—that laid waste to the wildness of the United States have been turned against the big opens of southern Canada and northern Mexico. During the last twenty years, Mexican cattlemen have waged a relentless campaign with 1080 poison against the remaining Lobos and Mexican Grizzlies. The Griz may be gone; the Lobo Wolves are a handful. Faced with the inexorable population growth and concomitant development of Mexico, the Jaguar and Ocelot have slipped farther south to their last stronghold in the disappearing rain forests of Central America and the Amazon.

Excepting the tropical rain forests, perhaps nowhere else is the war against the natural being waged so totally as in "Super Natural" British Columbia. In the summer of 1986, I spent a week with my wife's family at a fishing camp three hundred miles north of the U.S. border in British Columbia. Twenty years ago, it was a remote wilderness lodge surrounded by old-growth forests unending to the taiga, the Pacific Coast, the Great Plains, and the U.S. border. Today a constricting network of logging roads and huge clearcuts is choking the wildness from the land, unraveling the fabric that supports Grizzly, Wolverine, Fisher, Moose, Caribou, Lynx, and Gray Wolf (and British Columbia's economically important recreational fishing business). The British Columbian timber industry and its lapdogs in Parliament are creating a three-hundred-mile-wide swath of destruction through southern British Columbia.

That is why there is a Gray Wolf pack in Glacier; why there are more Grizzlies in the North Cascades; why Lobos, Jaguars, and Ocelots are being seen in Big Bend, the San Pedro, and the Peloncillos. Before, visits were occasionally made to the borderlands of the United States from population cores in Mexico or Canada; but today, as bad as the clearcutting, road building, overgrazing, and poaching are in the States, life is better in Montana, Washington, Arizona, New Mexico, and Texas than in British Columbia, Alberta, Sonora, Chihuahua, Durango, and Coahuila. Now they are coming for more visits—and to stay—in order to escape the devastation and persecution at home.

The refugia to the north and the south are no more. The wild country of the American West must stand on its own as a habitat for big bears, spotted cats, and wolves. It is the job of conservationists in the West to ensure that the habitat exists.

11 | The Principles of Monkeywrenching

I say, break the law.

—Henry David Thoreau

O nly 150 years ago, the Great Plains were a vast, waving sea of grass stretching from the Chihuahuan Desert of Mexico to the Boreal Forest of Canada, from the oak-hickory forests of the Ozarks to the Rocky Mountains. Bison blanketed the plains—it has been estimated that 60 million of the huge, shaggy beasts moved across the grassy ocean in seasonal migrations. Throngs of Pronghorn and Elk also filled this Pleistocene landscape. Packs of Gray Wolves and numerous Grizzly Bears followed the tremendous herds.

In 1830, John James Audubon sat on the banks of the Ohio River for three days as a single flock of Passenger Pigeons darkened the sky from horizon to horizon. He estimated there were several billion birds in that horde.

At the time of the Lewis and Clark Expedition, an estimated 100,000 Grizzly Bears roamed the western half of what is now the United States. The howl of the wolf was ubiquitous. The California Condor sailed the sky from the Pacific Coast to the Great Plains. Salmon and sturgeon populated the rivers. Ocelots, Jaguars, and

Jaguarundis prowled the Texas brush and the Southwestern mountains and mesas. Bighorn Sheep ranged the mountains of the Rockies, the Great Basin, the Southwest, and the Pacific Coast. Ivory-billed Woodpeckers and Carolina Parakeets filled the steamy forests of the Deep South. The land was alive.

East of the Mississippi, giant Tulip Poplars, American Chestnuts, oaks, hickories, and other trees formed the most diverse temperate deciduous forest in the world. In New England, White Pines grew to heights rivaling the Brobdignagian conifers of the far West. On the Pacific Coast, redwood, hemlock, Douglas-fir, spruce, cedar, fir, and pine formed the grandest forest on Earth.

In the space of a few generations we have laid waste to paradise. The Tall Grass Prairie has been transformed into a corn factory where wildlife means the exotic pheasant. The Short Grass Prairie is a grid of carefully fenced cow pastures and wheatfields. The Passenger Pigeon is no more; the last one died in the Cincinnati Zoo in 1914. The endless forests of the East are tame woodlots. The only virgin deciduous forest there is in tiny museum pieces of hundreds of acres. Fewer than one thousand Grizzlies remain. The last twenty-five condors are in zoos. Except in northern Minnesota and in Montana's Glacier National Park, Gray Wolves are known merely as scattered individuals drifting across the Canadian and perhaps the Mexican borders. Four percent of the Coast Redwood forest remains, and the ancient forests of Oregon are all but gone. The tropical cats have been shot and poisoned from our Southwestern borderlands. The subtropical Eden of Florida has been transmogrified into hotels and citrus orchards. Domestic cattle have grazed bare and radically altered the composition of the grassland communities of the West, displacing Elk, Bighorn Sheep, and Pronghorn, and leading to the virtual extermination of Grizzly, Gray Wolf, Cougar, and other "varmints." Dams choke most of the continent's rivers and streams.

Nonetheless, wildness and natural diversity remain. There are a few scattered grasslands ungrazed, stretches of free-flowing river, thousand-year-old forests, Eastern woodlands growing back to forest and reclaiming past roads, Grizzlies and wolves and lions and

Wolverines and Bighorn and Moose roaming the backcountry; hundreds of square miles that have never known the imprint of a tire, the bite of a drill, the rip of a 'dozer, the cut of a saw, the smell of gasoline.

These are the places that hold North America together, that contain the genetic information of life, that represent the eye of sanity in a whirlwind of madness.

In January 1979, the Forest Service announced the results of RARE II, the second Roadless Area Review and Evaluation: Of 80 million acres of undeveloped lands in the National Forests, only 15 million acres were recommended for protection against logging, road building, and other developments. In the big-tree state of Oregon, for example, only 370,000 acres were proposed for Wilderness protection out of 4.5 million acres of roadless, uncut forests. Of the areas nationally slated for preservation, most were too high, too dry, too cold, or too steep to offer much in the way of "resources" to loggers, miners, and graziers. Roadless areas with critical old-growth forests were allocated to the sawmill. Important Grizzly habitat in the Northern Rockies was tossed to the oil industry and the loggers. Off-road-vehicle dervishes and the landed gentry of the livestock industry won in the Southwest and the Great Basin.

During the early 1980s, the Forest Service developed its DARN (Development Activities in Roadless Non-selected) list, outlining specific projects in particular roadless areas. DARN's implications are staggering. The list is evidence that the leadership of the United States Forest Service deliberately sat down and asked themselves, "How can we keep from being plagued by conservationists and their damned Wilderness proposals? How can we ensure that we'll never have to do another RARE?" Their solution was simple: Get rid of the roadless areas. In its earliest form, DARN outlined nine thousand miles of road, one and a half million acres of timber cuts, and 7 million acres of oil and gas leases in National Forest RARE II areas before 1987. More recent figures from the Forest Service are far more disturbing: *The agency plans over half a million miles of new road, and up to 100,000 miles of this will be in roadless areas!* In most cases the damaged acreage will be far greater than the acreage stated, because the roads are designed to split undeveloped areas in half, and timber sales are engineered to take place in the center of roadless areas, thereby devastating the biological integrity of the larger area. The great roadless areas so critical to the maintenance of natural diversity will soon be gone. Species dependent on old growth and large wild areas will be shoved to the brink of extinction.

The Bureau of Land Management's Wilderness Review has been a similar process of attrition. It is unlikely that more than 9 million acres will be recommended for Wilderness out of the 60 million with which the review began. Again, it is the more scenic but biologically less rich areas that will be proposed for protection.

By 1990, Congress had passed legislation designating minimal National Forest Wilderness acreages for most states (generally only slightly larger than the pitiful RARE II recommendations and concentrating on "rocks and ice" instead of crucial forested lands). In the next few years, similarly picayune legislation for National Forest Wilderness in the remaining states (Montana and Idaho) and for BLM Wilderness will probably be enacted. The other roadless areas will be eliminated from consideration. Conventional means of protecting these millions of acres of wild country will largely dissipate. Judicial and administrative appeals for their protection are being closed off. Congress will turn a deaf ear to requests for additional Wildernesses so soon after disposing of the thorny issue. Political lobbying by conservation groups to protect endangered wildlands will cease to be effective. And in a decade, the saw, 'dozer, and drill will devastate most of what is unprotected. The battle for wilderness will be over. Perhaps 3 percent of the United States will be more or less protected, and it will be open season on the rest. Unless . . .

Many of the projects that will destroy roadless areas are economically marginal. For example, some Forest Service employees say that the construction costs for a low figure of 35,000 miles of roads in currently roadless areas will exceed $3 billion, while the timber to which they will provide access is worth less than $500 million. It is costly for the Forest Service, the BLM, timber companies, oil companies, mining companies, and others to scratch out the "resources" in these last wild areas. It is expensive to maintain the necessary infrastructure of roads for the exploitation of wildlands. The cost of repairs, the hassle, the delay, and the downtime may just be too much for the bureaucrats and exploiters to accept if a widely dispersed, unorganized, strategic movement of resistance spreads across the land.

It is time for women and men, individually and in small groups, to act heroically in defense of the wild, to put a monkeywrench into the gears of the machine that is destroying natural diversity. Though illegal, this strategic monkeywrenching can be safe, easy, fun, and—most important—effective in stopping timber cutting, road building, overgrazing, oil and gas exploration, mining, dam building, power-line construction, off-road-vehicle use, trapping, ski area development, and other forms of destruction of the wilderness, as well as cancerous suburban sprawl.

But it must be strategic, it must be thoughtful, it must be deliberate in order to succeed. Such a campaign of resistance would adhere to the following principles.

Monkeywrenching is nonviolent. Monkeywrenching is nonviolent resistance to the destruction of natural diversity and wilderness. It is never directed toward harming human beings or other forms of life. It is aimed at inanimate machines and tools that are destroying life. Care is always taken to minimize any possible threat to people, including the monkeywrenchers themselves.

Monkeywrenching is not organized. There can be no central direction or organization to monkeywrenching. Any type of network would invite infiltration, agents provocateurs, and repression. It is truly individual action. Because of this, communication among monkeywrenchers is difficult and dangerous. Anonymous discussion through the book *Ecodefense* and its future supplements seems to be the safest avenue of communication to refine techniques, security procedures, and strategy.

Monkeywrenching is individual. Monkeywrenching is done by individuals or very small groups of people who have known each other for years. Trust and a good working relationship are essential in such groups. The more people involved, the greater the dangers of infiltration or a loose mouth. Monkeywrenchers avoid working with people they haven't known for a long time, those who can't keep their mouths closed, and those with grandiose or violent ideas (they may be police agents or dangerous crackpots).

Monkeywrenching is targeted. Ecodefenders pick their targets. Mindless, erratic vandalism is counterproductive as well as unethical. Monkeywrenchers know that they do not stop a specific logging sale by destroying any piece of logging equipment they come across. They make sure it belongs to the proper culprit. They ask themselves what is the most vulnerable point of a wilderness-destroying project, and strike there. Senseless vandalism leads to loss of popular sympathy.

Monkeywrenching is timely. There are proper times and places for monkeywrenching. There are also times when monkey-wrenching may be counterproductive. Monkeywrenchers generally should not act when there is a nonviolent civil disobedience action—e.g., a blockade—taking place against the opposed project. Monkeywrenching may cloud the issue of direct action, and the blockaders could be blamed for the ecotage and be put in danger from the work crew or police. Blockades and monkeywrenching usually do not mix. Monkeywrenching may also be inappropriate when delicate political negotiations are taking place for the protection of a certain area. The Earth warrior always asks, Will monkey-wrenching help or hinder the protection of this place?

Monkeywrenching is dispersed. Monkeywrenching is a widespread movement across the United States. Government agencies and wilderness despoilers from Maine to Hawaii know that their destruction of natural diversity may meet resistance. Nationwide monkeywrenching will hasten overall industrial retreat from wild areas.

Monkeywrenching is diverse. All kinds of people, in all kinds of situations, can be monkeywrenchers. Some pick a large area of wild country, declare it wilderness in their own minds, and resist any intrusion into it. Others specialize against logging or off-road vehicles in a variety of areas. Certain monkeywrenchers may target a specific project, such as a giant power line, a road under construction, or an oil operation. Some operate in their backyards, while others lie low at home and plan their ecotage a thousand miles away. Some are loners, and others operate in small groups. Even Republicans monkeywrench.

Monkeywrenching is fun. Although it is serious and potentially dangerous activity, monkeywrenching is also fun. There is a rush of excitement, a sense of accomplishment, and unparalleled camaraderie from creeping about in the night resisting those "alien forces from Houston, Tokyo, Washington, D.C., and the Pentagon." As Ed Abbey said, "Enjoy, shipmates, enjoy."

Monkeywrenching is not revolutionary. It does not aim to overthrow any social, political, or economic system. It is merely nonviolent self-defense of the wild. It is aimed at keeping industrial "civilization" out of natural areas and causing its retreat from areas that should be wild. It is not major industrial sabotage. Explosives, firearms, and other dangerous tools are usually avoided; they invite greater scrutiny from law enforcement agencies, repression, and loss of public support.

Monkeywrenching is simple. The simplest possible tool is used. The safest tactic is employed. Elaborate commando operations are generally avoided. The most effective means for stopping the destruction of the wild are often the simplest. There are times when more detailed and complicated operations are necessary. But the monkeywrencher asks, What is the simplest way to do this?

Monkeywrenching is deliberate and ethical. Monkeywrenchers are very conscious of the gravity of what they do. They are deliberate about taking such a serious step. They are thoughtful, not cavalier. Monkeywrenchers—although nonviolent—are warriors.

They are exposing themselves to possible arrest or injury. It is not a casual or flippant affair. They keep a pure heart and mind about it. They remember that they are engaged in the most moral of all actions: protecting life, defending Earth.

A movement based on the above principles could protect millions of acres of wilderness more stringently than could any congressional act, could insure the propagation of the Grizzly and other threatened life forms better than could an army of game wardens, and could lead to the retreat of industrial civilization from large areas of forest, mountain, desert, prairie, seashore, swamp, tundra, and woodland that are better suited to the maintenance of native diversity than to the production of raw materials for overconsumptive technological human society.

If logging firms know that a timber sale is spiked, they won't bid on the timber. If a forest supervisor knows that a road will be continually destroyed, he won't try to build it. If seismographers know that they will be constantly harassed in an area, they won't go there. If ORVers know that they'll get flat tires miles from nowhere, they won't drive in such areas.

John Muir said that if it ever came to a war between the races, he would side with the bears. That day has arrived.

12 | In Defense of Monkeywrenching

At some point we must draw a line across the ground of our home and our being, drive a spear into the land, and say to the bulldozers, earthmovers, government and corporations, "thus far and no farther." If we do not, we shall later feel, instead of pride, the regret of Thoreau, that good but overly-bookish man, who wrote, near the end of his life, "If I repent of anything it is likely to be my good behavior."

—Edward Abbey

To no one's surprise, the advocacy and practice of monkeywrenching, or ecological sabotage, has become increasingly controversial during the last decade. Criticism of monkeywrenching has come from all corners of the political parking lot, from representatives of the timber, mining, and grazing industries and from mainstream conservation leaders and peace activists. It has come from politicians like timber-industry pit bull Senator James McClure of Idaho, as well as from some who claim to be conservationists, such as Representative Jolene Unsoeld of Washington; from right-wing proponents of the Genesis charge to "subdue" Earth, and from left-wing defenders of "the workingman."

Sometimes, when I hear public statements about monkeywrenching, I feel like a Coyote strolling through a Texas cow town and I tuck my tail between my legs and drop my ears low on my head and make tracks to the hills as fast as I can. Surprisingly often, though, when I'm expecting a load of buckshot or at least a hard-pitched rock in the ribs, I catch a wink and a corner-of-the-mouth smile and know someone's putting out leftovers on the back porch. It's important, in this funny place called America, to differentiate between what is said publicly and what dark heresies lurk in the quiet behind a pair of eyeballs. Support for monkeywrenching, despite all the proper cussing around the regulars' table in the town café, cuts across political, economic, educational, and other demographic lines. I know of realtors, Republican officeholders, septuagenarians, doctors, nurses, clergymen, college professors, construction workers, loggers, surveyors, park rangers, Forest Service employees, millionaires, lawyers, law-enforcement officers, grandmothers, and mainstream conservation group staffers who have monkeywrenched.

Monkeywrenching is a proud American tradition, existing happily in the shadows while decorous Americans bow before the brightly lit Great God Private Property. We all nod gravely to the town fathers and *grandes dames* outside of church, but many of us stifle a chuckle when Tom Sawyer lets the air out of their tires.

Monkeywrenching, ecological sabotage, ecotage, ecodefense, or "night work"—these are all terms for the destruction of machines or property that are used to destroy the natural world. Monkeywrenching includes such acts as pulling up survey stakes, putting sand in the crankcases of bulldozers, rendering dirt roads in wild areas impassable to vehicles, cutting down billboards, and removing and destroying trap lines. As noted in the preceding chapter, monkeywrenching is nonviolent and is aimed only at inanimate objects, *never* toward physically hurting people. It is defensive in that it is used to prevent destructive development in wild places and in seminatural areas next to cities. It is not major industrial sabotage; it is not revolutionary. Ecotage is not necessarily the most important tool for conservationists; it is merely one of many approaches that may be valid and effective, depending on the circumstances. The goals of monkeywrenching are to block environmentally destructive projects, to increase the costs of such projects and thereby make them economically unattractive, and to raise public awareness of the taxpayer-subsidized devastation of biological diversity occurring throughout the world.

Monkeywrenching, contrary to public opinion, is not a recent fad resulting from Edward Abbey's 1975 novel *The Monkey Wrench Gang,* or from my 1985 book *Ecodefense: A Field Guide to Monkey-wrenching.* It has a long and distinguished history in the United States, going all the way back to the Boston Tea Party, in 1773. There, good citizens of Boston, opposed to King George III's tyranny, boarded ships of the East India Company and dumped bales of tea worth hundreds of thousands of dollars into Boston harbor. Lester Rhodes points out in the *Earth First! Journal* that the modern practice of sabotaging private property to protect wilderness ecosystems has a direct precursor in the activities of the Underground Railroad before the Civil War, when citizens consciously broke the law by helping slaves escape into northern states and Canada.

Since at least as far back as the 1950s, preservationists have been cutting down billboards, pulling up survey stakes, decommissioning bulldozers, and committing other now-common acts of ecological sabotage. Abbey based *The Monkey Wrench Gang* on the real-life operations of several groups and individuals in the Southwest during the late 1960s and early 1970s. By 1985, when *Ecodefense* was published, ecotage was widespread. *Ecodefense* was published partly in an attempt to establish guidelines for monkeywrenching that would help it be more effective, strategic, ethical, safe, and secure.

Defenses of the legitimacy of monkeywrenching are varied, and depend upon the defender's general worldview and value system. For those whose values center on human beings and economics, who see the world as a storehouse of resources for the use of humans, who see no intrinsic value in nature, there may be no effective argument for ecodefense. (Inconsistencies do, however, stagger around our skull boxes like winos behind liquor stores. Even a small-town John Bircher might monkeywrench to protect his backyard or his favorite fishing hole.) Some critics genuinely value wild nature, but also believe in the supremacy of the law. While condemnations of monkeywrenching receive wide coverage and dutiful nods, I believe that many Americans can be convinced of the necessity of thoughtful ecotage as a last resort or where other methods are ineffective.

Nevertheless, those who practice or support ecotage should not lightly dismiss arguments against monkeywrenching, even though there is ultimately no resolution on some matters between persons holding diametrically opposed worldviews and value systems. The most difficult objections to monkeywrenching come from conservationists who view it as counterproductive to the preservation of the natural world. Because the ecodefender treads in a shadowy area of society, she needs to periodically question her motivations and review the effects of her actions. We should confront the diverse objections forthrightly. The longer you leave a crawl space open under your double-wide, the more skunks are going to hole up there.

The arguments against monkeywrenching, and my responses, include the following:

Current resource management is sound; there is no biological crisis; so monkeywrenching is unnecessary. This claim—from the logging, mining, and grazing industries and their employees; from some users of the out-of-doors like dirt bikers, trappers, and pickup-bound hunters; from the Forest Service and other "resource management" agencies; and from politicians pandering to rape-and-scrape industries—is a statement of faith, ignorance, or both. It is exemplified by a crowd of loggers who, during a July 1989 rally in Forks, Washington ("the Logging Capital of the World"), began chanting a mantra of "No! No! No!" to a speaker's question, "Do you think Spotted Owls can survive in old growth only?" Forget the research conducted by the Forest Service's own ecologists that conclusively demonstrates that Spotted Owls are old-growth-dependent. Accept on faith the anthropocentric doctrine that natural ecosystems are always improved by human management.

Throughout this book I offer evidence that, because of modern "resource management," Earth's biological diversity is severely threatened. Numerous other books and articles, many from leading scientists, concur. Sadly, the folly of believing all is right in the midst of a collapsing ecosystem has shaped history in many places, in many times. In Greece, 2,500 years ago, some pooh-poohed Plato's warning about the deforestation of that rocky but once-lush peninsula. A mere sixscore years ago in the United States, many frontiersmen saw no end to the buffalo.

The destruction of property is wrong. Here is a conflict of values. Those who support ecological sabotage in principle hold biological diversity and life in higher regard than they do inanimate private property. We realize that by damaging private property we are affecting the lives of those who own the property, as well as the jobs of those who operate the equipment. Nonetheless, it boils down to the question of whether private property (and those dollars or jobs the property represents) or natural ecosystems are more valuable. Although most people in this country (myself included) respect the concept of private property, life—the biological diversity of this planet—is far more important.

Breaking the law is wrong. There is not much room for negotiation between those who object, in principle, to breaking the laws of the state and those who believe that when higher values conflict with the laws of a political entity, one should break the laws. However, ecoteurs should carefully listen to this objection and not dismiss it out of hand. The law should *not* be broken for light or transient reasons.

If those who object to breaking the law under all circumstances were consistent, they would condemn the farmers of Lexington and Concord in 1775, as well as the demonstrators in Tiananmen Square in Beijing (which Henry Kissinger did, good ol' apologist for the supremacy of the state that he is), the Solidarity movement in Poland, Andrei Sakharov in the Soviet Union, the people of Romania who overthrew Ceausescu, and Nelson Mandela of South Africa.

Let us also remember that all of the resistance movements operating against the Nazis in Europe during World War II were acting illegally. Conversely, the Nazi atrocities were so completely legal that the allies had to write new laws hastily after World War II in order to prosecute Nazi war criminals at Nuremberg. There is obviously a difference between morality and the statutes of a government in power.

Monkeywrenching is undemocratic, and imposes extremist beliefs on others. An example of this argument would be, "The continued logging of ancient forests has been decided democratically through our political process. Once such a decision has been made, no one has the right to continue to oppose it, especially from outside of the law." This ignores evidence that our system is far from democratic—owing to the excessive power wielded by wealthy corporations to influence politicians through campaign donations and outright bribes, and through their advertising dollars in the media. It also ignores the fact that conservation groups, using the constitutional political process, have at times successfully halted timber sales by lawsuits in federal courts, only to have a few congressmen push legislation through Congress—without public hearings, committee hearings, or extensive debate—that prohibits citizens from challenging logging sales in the courts. Bureaucracies like the United States Forest Service are inherently undemocratic in promoting their own interests and their independence from public control.

Monkeywrenching is a tactic of poor losers. This argument is closely related to the previous one. Its proponents say it is poor sportsmanship to continue to try to protect roadless areas or other natural areas, once one's efforts to protect such areas have lost through the political process. Again, I answer that the political process is stacked in favor of those who are gobbling up fragile natural areas for a fast buck. It must also be remembered that a preservation victory by conservationists is always temporary: A "saved" river can be later dammed, a "saved" forest can be later cut, while a dammed river or a clearcut forest is almost irretrievably lost (at least in a human time scale).

We are not talking about a football game or a high-school debate here; we are discussing the continuation of three and half billion years of evolution. We are talking about what the ecologist Raymond Dasmann refers to as "World War Three"—the war between humans and Earth.

Monkeywrenching isn't playing fair. Falling on the heels of the previous two arguments, this one attests to the effectiveness of monkeywrenching. The answer to it is a question: When have corporations ever played fair? Corporations establish the rules of the political game to favor themselves, and when citizens manage to achieve success, those with economic power go outside the rules or change the rules. A case in point is the proposed University of Arizona astronomy complex on Mount Graham in Arizona. Virtually every reputable biologist who has studied this project agrees that it would cause extensive damage to a relict ice-age spruce-fir forest and possible extinction of the Endangered Mount Graham Red Squirrel. Moreover, there are alternative sites just as good or better for astronomy.

In 1988, mainstream conservation groups appeared to be about to kill the project through the required environmental impact statement and Endangered Species Act review. Thereupon, the University of Arizona, desiring the prestige and funding of the project and operating like an economically powerful corporation, called in its political chits with the Arizona congressional delegation. The delegation, including even supposed environmental godfather Representative Morris Udall, pushed through Congress an end-run on both the Endangered Species Act and the National Environmental Policy Act that decreed the building of the observatory. The only options left to conservationists to protect both Mount Graham and the sidelined environmental laws became civil disobedience and monkeywrenching (although conservation groups filed a long-shot suit as well).

Of course, the call for "fair play" hearkens back to the American Revolution, when the British complained that the American rebels didn't fight fair because they hid behind trees to shoot, and didn't stand out in the open, like men. The "lobsterbacks" called Washington's troops cowards, just as loggers today call monkeywrenchers cowards.

Monkeywrenching threatens human beings with injury or death; it is "eco-terrorism." Demagogues, whether editorial writers, politicians, or industry spokespersons, have found that this hysterical claim gets the most attention. This argument is directed most often against tree spiking, although it is also used against other forms of ecotage. The monkeywrencher must very carefully weigh the possibility of harm to a person from any ecotage, and must act to insure that no humans are hurt. *Ecodefense* constantly underscores this point and offers many safety suggestions—including the need for clear, direct warnings about spiked trees. (I discuss tree spiking in more detail in the next chapter. It consists of hammering large nails that can damage saws into trees slated to be logged.) Contrary to the demagogues' claims, no one has been injured in any monkeywrenching operation carried out by preservationists.

The true ecoterrorists are the planet-despoilers: Those in the Forest Service and the timber industry who are annihilating thousand-year-old forests for paper bags and picnic tables. Ranchers and employees of the Department of Agriculture's Animal Damage Control unit who have exterminated predators ranging from Grizzly Bears and Gray Wolves to Common Ravens and Bobcats and continue to slaughter them in their remnant ranges. The calculator-rational engineers and pork-barrel politicians who want to plug every free-flowing river with dams. The thrill-seeking dirt bikers who terrify wildlife and scar delicate watersheds with mindless play. Japanese and Icelandic whalers who are hounding the last great whales to the ends of the Earth, despite international agreements against whaling. The heads of Exxon and other giant oil companies who cut back on safety measures to save a few pennies and thereby cause disasters like the Prince William Sound oil-tanker wreck and the blowout of drilling platforms in the Santa Barbara Channel. Corporate executives whose bottom line is profit and who could not care less about Love Canals, Bhopals, cigarette smoke, acid rain, and unsafe automobiles. Otherworldly "religious leaders" who condemn birth control and encourage the poor in Third World countries to have more children. The list of ecoterrorists is endless—but it does not include the brave and conscientious individuals who are defending threatened wild areas by placing a monkeywrench into the gears of the machine.

Ignored in the debate about ecoterrorism are the simple facts about who has been hurt by whom. Conservationists had to attend northern California hearings on the Redwood National Park expansion in the 1970s under police escort, because of threats of violence by loggers. A pair of anti-wilderness ranchers attempted to push politically moderate conservationist Dick Carter over a cliff to his death on a BLM field trip in southern Utah in the 1970s. Vehicles with Sierra Club bumper stickers were trashed at Escalante Canyon trailheads in the mid-1970s by locals supporting a massive coal-fired power plant on southern Utah's Kaiparowits Mesa, which the Club was fighting (and defeated). As the Southwest representative for The Wilderness Society in the 1970s, I had my life threatened twice by residents of Glenwood, New Mexico, where I lived at the time— once over legislation to study the San Francisco River for Wild and Scenic River status, and once over RARE II. Advocates of massive dirt-bike races in the California Desert Conservation Area attended BLM hearings in the early 1980s wearing pistols to intimidate conservationists and BLM employees. Denzel and Nancy Ferguson, managers of a university field station on the Malheur National Wildlife Refuge in eastern Oregon, were roughed up and thrown out of a community dance by five ranchers after they complained about overgrazing on the Refuge. Buzz Youens, a retired hospital architect and a vocal and effective opponent of logging on the Apache National Forest in eastern Arizona, disappeared in 1978 after threats on his life by loggers. His decaying body was found handcuffed to a tree more than a year later. He had been shot.

After nonviolent protests against strip-mining of pumice for stone-washed blue jeans in the Santa Fe National Forest in 1989, an individual wearing an Earth First! T-shirt was beaten by patrons of a bar he unfortunately entered in Jemez Springs, New Mexico. A Forest Service law-enforcement officer, Billy Ball, cut down a tree in which an Earth First! activist, James Jackson, was perched to protest the destruction of the proposed Four Notch Wilderness Area in east Texas. Jackson's legs were permanently injured. Billy Ball later threatened another EF!er, Barbara Dugelby, saying he would have the IRS audit her parents' tax returns if she didn't call off a protest she was organizing. Jeff Elliott, a junior-high-school biology teacher, had his logging-town home burned to the ground after he became active with New Hampshire Earth First! Four Earth First! women arrested after occupying a logging yarder* in Oregon's Siskiyou National Forest were beaten by local women in their cell because of the logging controversy and because one of the EF! women was black. In 1983, a bulldozer operator working on the Bald Mountain logging road in the Siskiyou National Forest buried five blockaders in dirt. A few days later, as a man in a wheelchair and I peacefully blocked the road, Les Moore, the driver of a truck bringing five bulldozer jockeys to the work site, ran over me and pushed me one hundred yards with his truck.

In 1989, the driver of a logging truck rammed a small car driven by Earth First! organizer Judi Bari as she drove down a logging road in northern California. He mumbled that he didn't know kids were in the car, as Bari and four children crawled out of the wreckage. In 1990, Bari and other organizers of the "Mississippi Summer in the Redwoods" protests began receiving crude death threats in the mail and nailed to the door of the Mendocino Environmental Center, in Ukiah. Authorities, in a frightening refrain of the original Mississippi Summer, refused to investigate or prosecute. Bari was almost killed in late May 1990 when a pipe bomb exploded in her car. Oakland city police arrested her on trumped-up charges of transporting a bomb, and failed to seek the real bomber. Later, the

*A logging yarder is a large crane-like machine that uses cables to pull cut logs into a yard where they can be loaded onto trucks.

District Attorney declined to press charges, but no real effort has been made by the FBI or police to find the bomber.

These are a handful of the incidents of physical violence directed toward conservationists over the last twenty years. Where are the ecoterrorists, indeed?

Finally, let us consider the statements of government officials about terrorism. FBI agent David Small defined terrorism at a Phoenix, Arizona, press conference following the arrests of Mark Davis, Marc Baker, Peg Millett, and me on May 31, 1989. Terrorism, he declared, was the breaking of the law for political or social reasons. Under this definition, Reverend Martin Luther King, Jr., Rosa Parks, Mohandas K. Gandhi, Henry David Thoreau, George Washington, Betsy Ross, Thomas Jefferson, and Joan of Arc were all terrorists.

In contrast to Small's definition, former FBI Director William Webster had a much stricter view of what constitutes terrorism. In the December 21, 1984, *New York Times,* Dr. Warren M. Hern reported:

> *Two dozen abortion clinics or offices have been bombed, burned or otherwise attacked this year. Some have been destroyed. . . . One physician was kidnapped with his wife and held hostage for a week. . . . Another doctor's wife and child narrowly escaped injury when rifle bullets riddled their house. . . .*
>
> *William H. Webster, Director of the Federal Bureau of Investigation, recently stated that the attacks on abortion clinics do not constitute "terrorism" because they were not committed by an "organized group" that publicly took responsibility for the attacks. The FBI, he said, only investigates "true terrorism" that aims to "overthrow the Government" or "shift the Government"; attacks on abortion clinics—not politically motivated—are "low priority."*

Of course, the Reagan administration warmly embraced antiabortion zealots while condemning mainstream conservationists, like Sierra Club members, as "environmental extremists." This may explain why the FBI under Reagan ignored the violence directed toward people by anti-abortionists, and why the FBI, under Reagan and, later, under Bush, spent over $2 million in an attempt to frame me for "conspiracy" and brand the Earth First! movement as "terrorists."

In 1976, President Jorge Rafael Videla of Argentina said, "A terrorist is not just someone with a gun or a bomb, but also someone who spreads ideas that are contrary to Western and Christian civilization." Many in government and industry in the United States would agree with that definition. One only needs to consider the legacy of the Argentine generals to appreciate the horrors to which that attitude leads.

Monkeywrenching turns one into a criminal. In the narrow eyes of the law, this is true. Anyone who contemplates pulling up survey stakes or decommissioning a bulldozer should carefully think about her decision and prepare herself for the possible consequences—which include everything from damaged reputation, imprisonment, and fines to actual physical harm. Wilderness guide and Earth First! cofounder Howie Wolke, while pulling up survey stakes for a planned oil exploration road into fragile Elk habitat in Wyoming, was assaulted by a surveyor with a hatchet and could have been seriously injured. The local sheriff refused to consider charges of assault against the surveyor, and Howie was convicted of a misdemeanor and sentenced to the maximum: six months in a cramped county jail cell. Fortunately, very few monkeywrenchers have ever been arrested.

When will the real criminals go to jail? Why is the CEO of Exxon Corporation not in a federal penitentiary for cutting back on safety measures and causing the catastrophic oil spill in Alaska's Prince William Sound? Why is industrialist Armand Hammer not doing life without parole for the Love Canal murders? Why has not India invaded the United States to capture the top executives of Union Carbide to make them stand trial for mass murder in Bhopal?

Monkeywrenching leads to vandalism, disrespect for law, and the breakdown of society. A committed and serious monkeywrencher has high moral standards. Indeed, such ethics are inherent to appropriate ecotage. However, the potential for descent into thoughtless vandalism is a troubling argument against monkeywrenching. Anyone who chooses to protect the wild extralegally, by destroying property, needs to guard against becoming a mere hooligan. Advocates of monkeywrenching should take care not to publicly endorse specific acts of foolish or irresponsible vandalism posing as ecotage.

By advocating monkeywrenching, one is liable for all excesses—even those where someone has obviously strayed from published guidelines. This, too, raises thorny questions. Where does responsibility lie when something goes wrong? If, someday, someone is injured by improper or unpublicized tree spiking, will monkeywrenching advocates be responsible? Will they be responsible if the property of an innocent party is destroyed? As editor and publisher of *Ecodefense*, I am concerned about these questions. Regardless of how careful one is to offer the best information and the strongest encouragement to operate ethically and deliberately, there is no guarantee that every monkeywrencher will practice safety and discrimination. I know one highly principled, conscientious advocate of ecotage who no longer encourages it because he questions the likelihood that all monkeywrenchers will be as strategic-minded as he is.

I believe in individual responsibility, and I believe that I have acted responsibly in stressing safety, deliberateness, and strategic considerations in *Ecodefense*. I believe that each individual monkeywrencher has a similar responsibility when taking action.

Those who argue against providing ideas and tools to others because they may not act responsibly are profoundly antidemocratic. They fear placing power in the hands of the people, and think it should be held only by "authorities" or "institutions." History shows that although individuals have committed heinous crimes, institutions, authorities, and states have committed far worse. *Ecodefense* arms the people with tools to thwart injustice committed by the state and corporations against wild nature.

My experience is that monkeywrenchers are thoughtful and careful. With few exceptions, ecodefenders deliberately select proper targets, act strategically with the larger picture in mind, and are rigorous in being sure no one is hurt by their wilderness defense. The monkeywrenchers I have met impress me as highly ethical and conscientious individuals. Rare are cases of monkeywrenching by nihilists lashing out blindly at what they perceive as an unfair society. In such incidents, the vandals generally do not need *Ecodefense* as a field guide; they have their own techniques.

In most cases of which I am aware where unsafe techniques were used, or where the monkeywrenching was counterproductive in a strategic sense, or where an unjustifiable target was selected, there is good evidence that the sabotage was not done for ecological reasons, but that it was mere vandalism by kids out for kicks; that it was done by industry to give preservationists a bad name; or that it was done for insurance reimbursement by the owner of the property.

Nevertheless, the fact remains that it is possible for individuals to use published techniques of monkeywrenching for unwarrantable reasons. Even if the above considerations did not justify monkeywrenching, we need to balance the possibility of misuse against the severity of the crisis to which most ecodefenders are responding. We need to compare the potential harm done by strategically unwise monkeywrenching with the harm that would occur if responsible monkeywrenching stopped. The balance sheet solidly attests to the need for responsible monkeywrenching.

If you monkeywrench, you should take credit for your actions and accept punishment as one does for any act of civil disobedience. Overlooked here are the fundamental differences between civil disobedience and monkeywrenching. The goal of civil disobedience in most cases is to reform society or some aspect thereof by conscientiously and nonviolently violating the law (as in a blockade), thereby appealing to the public and reasonably fair authorities with the rightness of one's cause and personal integrity. In other cases, it is to witness against evil being done, to refuse to acquiesce to that evil, and thereby to grow spiritually. In both cases, arrest and punishment are integral elements of the action.[1]

Monkeywrenching, on the other hand, is aimed not at reforming society but at thwarting destruction. Although a similarly high level of deliberate and ethical behavior is required, spiritual growth is not a specific goal of ecotage (although it may be a side benefit). What is important is stopping the damage; the monkeywrencher, like the guerrilla fighter, is more effective when avoiding capture and being able to return again and again.

Monkeywrenching "lowers the standards of civility another notch." Columnist Alston Chase argues this, and further opines that "the popularity of direct action signals a declining faith in reason." While this objection should not be dismissed with a snort, one may well ask what reason has brought us. Has it brought us Love Canal, *Exxon Valdez*, Chernobyl, Three Mile Island, overpopulation, clearing of the tropical rain forests, the "incidental" slaughter of dolphins in tuna fishing, the destruction of the ozone layer, an epidemic of environmentally induced cancers, a United States dominated by a smaller and smaller cabal of corporate executives, and the poisoning of air and water with industrial by-products? Putting grinding compound into the crankcase of a logging yarder may not be civil, but crucifying three and half billion years of evolution on a cross of gold is far more uncivil. Eco-saboteurs have taken off the white gloves and are getting their fingernails dirty. It's a nasty job, but somebody's got to do it.

By using economic loss as a deterrent, monkeywrenchers buy into the materialistic system. This reasoning comes from a few pure souls in the counterculture. I would argue instead

1. Many Earth First!ers would argue that thwarting destructive projects is the purpose of the civil disobedience they commit. Although this has been my motivation when participating in civil disobedience, it is not the classic strategy of civil disobedience. CD that is goal-directed like this is actually more akin to monkeywrenching than to the civil disobedience practiced by Gandhi, the Civil Rights movement, and the pacifist or disarmament movement.

that monkeywrenching is like an Eastern martial art that turns an opponent's superior strength against himself. When outnumbered and outgunned, you look for your opponent's weak spot, the place where he does not want to be hit. For corporate America, that is generally the pocketbook.

Destroying machines is wrong. My friend Gary Snyder, a Pulitzer Prize–winning poet, has argued that we should respect tools and machines as complex, useful entities, almost akin to living systems. I agree that a good craftsperson should respect her tools and take care of them, but when we consider a bulldozer that is ripping up biological diversity, I think we need to take a Zen approach. What is a bulldozer made of? Metal, largely iron ore. And from where does that ore come? Earth. The bulldozer has been made out of the Earth. It doesn't want to be tearing around destroying Earth. The monkeywrencher is a bodhisattva performing an act of *satyagraha*, enabling the bulldozer to find its true dharma nature, its Buddhahood, by returning it to Earth as a lump of rusting metal.

Monkeywrenching costs jobs. This argument is primarily used by large timber companies to whip their gullible employees into a red-eyed stew against any conservationist tactic, including ecotage. Lost in the shuck and jive from the front-office PR flacks are the statistics. Although the timber harvest from the National Forests is at an all-time high, employment in harvesting that timber has plummeted in recent decades. Jeff DeBonis, a timber sale planner on Oregon's Willamette National Forest, reports:

> *Between 1979 and 1989 the timber harvest on federal lands in Oregon increased 18.5%, in that same period employment in the wood products industry dropped 15%. The claim by the timber industry that employment is tied to the harvest level on the national forests is simply not true. . . . The lumber and wood products industry now contributes only 6% to Oregon's total economy; in the last decades while employ-*

ment in the wood products industry dropped 15%, total employment in the state is up 24%. [2]

Logging and milling employment is down because large timber companies are interested in profits, not employees. Replacing men with machines results in higher profits; corporate executives then don't have to worry about disability benefits for injuries suffered owing to rushed and unsafe working conditions, or about machines going on strike. (They do have to worry about those machines getting monkeywrenched, though.)

Sabotaging and thereby closing down Hitler's extermination camps would have cost jobs, too (and it did). Locking up burglars and armed robbers takes away their chosen livelihood. However, some things are more important than jobs; Earth's remaining biological diversity is among them.

Monkeywrenching is ineffective. This is claimed, possibly as a face-saving effort, by some in industry, politics, and the Forest Service. It is also said by those mainstream conservationists who wish to discourage ecotage for various reasons. The effectiveness of monkeywrenching *is* difficult to gauge. Certainly not every project ever sabotaged has been stopped. Nor has a growing level of monkeywrenching yet forced the retreat of industrial civilization from wild areas (which I optimistically predicted in the first edition of *Ecodefense;* it may still occur, however).

Nevertheless, a number of incidents of ecotage have stopped destructive projects or have played a role with other tactics in stopping them. The crippling of the Icelandic whaling fleet in a daring raid by two members of the Sea Shepherd Conservation Society has sharply reduced Iceland's whale kill. A carefully planned and extensive tree-spiking operation in ancient cedars on British Columbia's Meares Island was a significant factor in preventing clearcutting planned there. Tree spikings in Washington, Oregon, Virginia,

2. Quoted from *Thunderbear,* an alternative newsletter for the National Park Service, issue #110 (Nov. 1989), New Orleans, LA.

New Mexico, and Montana are known to have stopped proposed timber sales. Many conservationists believe that ecotage, including tree spiking, was a significant factor in stopping the logging of Bowen Gulch in Colorado in 1990. Steady monkeywrenching over several years led to the retirement of cattle grazing in a desert region in Arizona. Raids by the Animal Liberation Front on animal-experimentation labs have led to greater restrictions on such projects and increased public opposition to animal torture for bogus science. Steady monkeywrenching of seismographic equipment hastened the cessation of oil and gas exploration around Jackson Hole. In 1985, ecoteurs firebombed a $250,000 chipper in Hawaii that was grinding up rain forest for power-plant fuel; the operation went bankrupt. Carefully targeted ecotage aided in stopping a major subdivision next to critical Bighorn Sheep range in the Catalina Mountains near Tucson in the 1970s; the area is now a protected state park. More generally, widespread ecotage has helped increase public opposition to logging ancient forests in the United States, Canada, and Australia. Monkeywrenching has become a particularly effective and widespread tool used by land-based people in Third World countries such as Thailand, Brazil, and Malaysia. Sabotage of polluting industries is more common in Europe than in North America.

One researcher tried to gauge the overall impact of monkeywrenching for the *Earth First! Journal* in 1990. He reports that Forest Service special agent Ben Hull surveyed forest supervisors across the nation, and afterward refused to discuss his findings because "he doesn't want to give ecoteurs the satisfaction of knowing how much havoc they're causing." The Association of Oregon Loggers believes that the average ecotage incident costs $60,000 in equipment loss and downtime. Law-enforcement and insurance costs drive this figure even higher. The *EF! Journal* investigator estimated that ecotage in the National Forests alone in the United States is costing industry and government $20–25 million annually.[3]

The greatest effect of monkeywrenching is a general discourage-

---

3. C.M., "An Appraisal of Monkeywrenching," *Earth First! Journal* February 2, 1990.

ment of projects destructive to nature. By carefully targeting logging equipment used against natural forests, ecodefenders drive up the cost of insurance for such operations. Some contractors nowadays refuse to bid on Forest Service road projects or timber sales in roadless areas because of the likelihood of damaged equipment and resulting cost overruns. Companies bidding on road construction for the Mount Graham Astronomical Observatory complex in Arizona doubled their bids because of fears of monkeywrenching and the need for round-the-clock security on site.

Monkeywrenching gives the entire environmental movement a bad name and causes an antienvironmental backlash. This may be the most complicated argument against monkeywrenching, and one of the most important to understand. In the larger picture, the backlash argument essentially says, "If your tactics or demands can be characterized as 'extremist,' you will besmirch those whose tactics or demands are more moderate, and you will cause a backlash that will overturn previous advances for your cause."

Dr. Martin Luther King, Jr., heard such warnings in 1955 when he began to organize civil-disobedience protests against discrimination and segregation in the South. Leaders of moderate civil-rights groups asked him not to rock the boat with excessive demands and confrontational tactics. They told him that although conditions were far from perfect, they were improved; if he asked for too much, or angered white racists, the modest gains since World War II would be rolled back. The threat of a backlash was also made by both the segregationist Southern establishment and racist poor white groups such as the Ku Klux Klan, and some carried out such threats against civil-rights workers with police brutality, murder, and assault. Fortunately, King and his compatriots had both a dream and the courage to pursue it. Had they listened to the nervous Nellies of the moderate groups and to white racists, measures like the 1964 Civil Rights Act and the 1965 Voting Rights Act might never have been achieved.

The labor movement was warned early in the century about

extremism tarnishing its acceptability in America. Here moderates prevailed, and became willing partners of management. The increasingly irrelevant state of unions is a result.

Radicals in the anti–Vietnam War movement were blamed for prolonging the war and for damaging the "respectable" opposition's appeal to the American public. However, there is good evidence that fear of increasingly militant demonstrations kept President Richard Nixon from using nuclear weapons in Vietnam, and that such stridency is what finally wore down the pro-war establishment.

I recall many times during my twenty years in the conservation business when arguments about possible backlash were trotted out to convince preservationists to moderate their demands. Throughout the 1970s, the timber industry used the threat of backlash to warn conservationists against including forested lands in Wilderness Area proposals. Dirt bikers used it to admonish conservationists for trying to restrict their free use of the public lands. The mining and oil and gas industries are masters of this argument—warning conservationists that the public will rebel if their "land lockup" policies drive up energy or mineral prices. James Watt and Ronald Reagan brandished it in the early 1980s when they called The Wilderness Society and other groups "environmental extremists." The so-called Sagebrush Rebellion was the greatest backlash hoax ever perpetrated; it was launched by the livestock industry to shore up their crumbling edifice on the Western public lands when conservationists finally began to argue for cutbacks in livestock.

During the last decade, Earth First! and monkeywrenching have received the lion's share of "environmental extremist" catcalls, warnings of backlashes, and allegations of tactics counterproductive to legitimate environmental aspirations. These threats arise from the antienvironmental lobby, Milquetoast environmentalists who are afraid of industry's shadow, and conservation leaders who covet respectability.

The conservation issue that most often engenders the threat of backlash today is the campaign to preserve the scattered remnants of ancient forests in the Pacific Northwest. Using this threat, the Oregon and Washington congressional delegations, led by Senator

Mark Hatfield, intimidated the Sierra Club into helping draft a legislative rider in the fall of 1989. This rider, which was passed into law without hearings or open discussion, denies citizens access to the courts to sue against timber sales, and mandates the largest timber harvest ever of old-growth public forests. Grassroots groups including the Oregon Natural Resources Council and the Native Forests Council opposed the Sierra Club sellout, as did the Sierra Club Legal Defense Fund, the National Audubon Society, and the National Wildlife Federation, but to no avail. I believe that future generations will judge the Sierra Club more harshly on this pitiful sellout of ancient forests than on any other single failure.

In previous years, similar threats of backlash prompted conservationists to delete timbered areas from Wilderness proposals, resulting in a National Wilderness Preservation System that emphasizes scenic "rocks and ice" suitable for backpacking, and excludes productive forests important for habitat and biological diversity. Even Earth First!ers are not immune to such insecurity. In April 1990, some northern California, Oregon, and Washington Earth First! groups renounced the tactic of tree spiking in a futile attempt to develop alliances with timber mill workers.

With this background, we can better weigh the backlash argument as applied to monkeywrenching. Even though it is a standard warning used by the political establishment to keep any viable challenge in line, it still should be carefully considered by proponents of night work. We need to mull over several questions:

■ Is there any evidence of monkeywrenching tarring environmentalism with a bad name?

■ Is there any evidence of a backlash against conservation caused by monkeywrenching?

■ If so, would the backlash have come anyway from certain groups (loggers, for example)?

■ Is the backlash more rhetorical than real?

■ Does the backlash (if it exists) outweigh the gains achieved by monkeywrenching?

There are certainly attempts to tar the environmental movement with the "bad name" of monkeywrenching by industry (particularly the logging industry), opportunistic politicians, certain editorialists, and antienvironment right-wingers. But there is little evidence that such attempts have succeeded. While ecotage is associated in the public mind with Earth First!, it has not been successfully attached to more moderate groups. (And here I speak from direct observation of attempts to do just this when mainstream groups such as the Idaho Conservation League, the Alaska Conservation Foundation, and the national Sierra Club have invited me to speak at their conventions.)

There may appear to be a backlash against environmental protection measures such as Wilderness Area designations and restrictions on timber harvest, but there is no convincing evidence that this is a reaction to monkeywrenching. In fact, closer analysis reveals that there is no antienvironmental backlash at all. For the twenty years I have been involved in the wilderness-preservation movement, there have been warnings of such a backlash, but the so-called backlash has never been anything more than the standard opposition to Wilderness designations or public-land protection from the timber, mining, energy, and grazing industries and from lowbrow recreationists like off-roaders and slob hunters. The only backlash that has actually occurred is the Earth First! movement, which happened when some once-moderate conservationists got tired of being stomped on by industry and rednecks.

The supposed antienvironmental backlash among loggers in the Northwest today is merely the imitation of conservationists' successful organizing techniques by a longtime anticonservation industry. While loggers find ecotage a handy target to blast, they are just as likely to choose Spotted Owls, Wilderness Areas, and National Parks as targets, and the myth that "old-growth forests are decadent and wasted" as their ammunition. This unthinking opposition to forest protection among loggers and their small-town peers has been fanned by the large timber corporations in order to fight a belatedly aroused conservation movement that is trying to protect remnant ancient forests.

In short, the "backlash against monkeywrenching" is more rhetorical than real. This does not mean that there is not real anger or fanaticism among loggers and others that may lead to violence—and we have already seen that it has led to violence—but that ecotage is just an appealing target rather than the root cause.

Given all of this, the potential for ecotage protecting wild areas that cannot be saved by standard conservation tools far outweighs the potential for giving the entire environmental movement a bad name and causing an effective backlash that involves portions of the public not viscerally involved in land-use-allocation decisions.

In the early 1980s, the *Grand Junction* [Colorado] *Sentinel* editorialized against Earth First! and urged the Sierra Club, which they praised, to sue EF! for damaging the reputation and image of conservation groups. They neglected to mention previous editorials criticizing the Sierra Club in equally harsh terms. A major accomplishment of Earth First! (and of ecotage) has been to expand the environmental spectrum to where the Sierra Club and other groups are perceived as moderates. This has more than offset any negative connotations arising from conservation's connections with monkeywrenching.

Let's move from defense to offense here, and consider some arguments *for* monkeywrenching. There are a number of basic arguments for ecological sabotage, several of which have solid precedents in American history and the ideas of Western civilization. You never know when you'll be called on to state your case.

Biophilia. The Harvard biologist E. O. Wilson argues that many defend nature because of *biophilia*, an inherent love for nature. Ecodefenders are motivated by a great love for wilderness, biological diversity, and Earth. Aldo Leopold wrote in the introduction to his conservation classic, *A Sand County Almanac*: "There are some who can live without wild things, and some who cannot. These essays are the delights and dilemmas of one who cannot."

Acts of monkeywrenching are the delights and dilemmas of those who cannot live without wild things, and who see their hands otherwise tied by a political system that serves an economic elite.

Self-defense. Aldo Leopold argued that we need to "think like a mountain," and Australian John Seed, architect of the international rain-forest preservation movement, argues in his essay "Beyond Anthropocentrism" that by identifying with the wilderness, one *becomes* the wilderness: "I am protecting the rain forest" develops to "I am part of the rain forest protecting myself. I am that part of the rain forest recently emerged into thinking."

When we fully identify with a wild place, then, monkeywrenching becomes self-defense, which is a fundamental right. It is important to note that this kind of self-defense is done in humility. The ecodefender is not a superior being protecting something less than herself, but is an antibody of the wildland acting in self-defense, and drawing on the forest or desert or sea for wisdom, strength, and strategy.

Defense of one's home. Edward Abbey develops this argument in his "Forward!" to *Ecodefense*:

> *If a stranger batters your door down with an axe, threatens your family and yourself with deadly weapons, and proceeds to loot your home of whatever he wants, he is committing what is universally recognized—by law and morality—as a crime. In such a situation the householder has both the right and the obligation to defend himself, his family, and his property by whatever means are necessary. This right and this obligation is universally recognized, justified and even praised by all civilized human communities. Self-defense against such attack is one of the basic laws not only of human society but of life itself, not only of human life but of all life.*
>
> *The American wilderness, what little remains, is now undergoing exactly such an assault. . . .*
>
> *For many of us, perhaps for most of us, the wilderness is as much our home, or a lot more so, than the wretched little stucco boxes, plywood apartments, and wallboard condominiums in which we are mostly confined by the insatiable demands of an overcrowded and ever-expanding industrial culture. And if the wilderness is our true home and if it is threatened with invasion, pillage and destruction—as it certainly is—then we have the right to defend that home, as we*

would our private rooms, by whatever means are necessary. (An Englishman's home is his castle; an American's home is his favorite fishing stream, his favorite mountain range, his favorite desert canyon, his favorite swamp or patch of woods or God-created lake.)

To obtain protection denied us by a government that ignores the will of the people. Public-opinion polls consistently demonstrate that the people want stronger protection of the environment—including clean air and water, control of toxics, protection of wilderness, forests, and wildlife—than the government will undertake. The reason for this is that our governments—local, state, and federal, including both elected officials and bureaucrats—are more influenced by corporations and a wealthy elite than by the people.

Furthermore, corporate America owns the electronic and print media, and thereby controls the presentation of the news. The advertising dollar makes some subjects taboo for news coverage and prevents presentation of effective critiques of corporate control. A recent example was a 1989 National Audubon Society television special about the logging of ancient forests in the Pacific Northwest. Logging corporations persuaded other corporations, including Ford Motors and Stroh's Beer, to pull their scheduled advertising off the program in an attempt to prevent its broadcast. Only because Ted Turner was financially able to present the program on his network without advertising and absorb the financial loss did the program air.

Similarly, federal agencies such as the Forest Service and the Army Corps of Engineers twist environmental-reform legislation like the National Environmental Policy Act (NEPA) to their own ends. After twenty years of dealing with these agencies on projects that require formal public input and the preparation of an environmental impact statement before a decision is supposedly reached, I have seen an unbreakable pattern emerge: *Instead of using the planning process or the environmental-impact study as a tool of analysis to guide the agency to a management decision, it is used as a paper trail to justify a previously made in-house decision; instead of seeing public*

involvement as a means to gain outside expertise, it is seen as something
to be manipulated so that it appears to support the agency decision.

There are several reasons for this usually conscious, but sometimes unconscious, perversion of the spirit of the law by bureaucracies. First, the agencies are run by professionals—"expert" foresters,
range managers, engineers, and the like. They think they know
what to do and don't need outside interference. Second, quasi-
regulatory agencies have historically slept with the industries they
are supposed to regulate. The interests of the industry become their
interests. Each agency identifies with its constituency: the Forest
Service with the timber industry; the BLM with the grazing and
mining industries; dam builders with large irrigators and local politicians feeding at the pork barrel; the Park Service with concessionaires and industrial tourists. Third, as the forest economist Ran-
dal O'Toole shows in his book *Reforming the Forest Service,*[4]
government agencies are motivated toward those programs that
bring them money, status, power, and growth. Of course, Wilderness Areas and other protected lands need less management and
fewer managers than lands open to all sorts of destructive "management" schemes.

So how should conservationists attempt to influence management
by the Forest Service and other agencies? To begin with, forget
about reforming the agencies. That's been tried with NEPA,
NFMA, RPA, FLPMA. . . . Reform fails because agencies are
supremely competent at subverting reform laws (this may be the
only thing at which they *are* competent). They can manipulate
different interest groups to their advantage. If everyone seems un-
happy with what they are doing, they claim they are successfully
balancing all interest groups and achieving the "greatest good for
the greatest number in the long run."

Because the government refuses to protect the land as the citizens
want, the people have the right to thwart the destruction of Earth

4. Washington, D.C.: Island Press, 1988. O'Toole, a brilliant forest economist
and noted gunfighter for hire by conservation groups, offers an incisive and provocative application of libertarian economics to management of the public lands. His
book deserves to be carefully read and considered by all conservationists.

through monkeywrenching, just as the citizens of Boston had the right to thwart the King of England and exploitative British companies with the Boston Tea Party.

To extend rights to those excluded from the law. Corporations, nonliving entities, are persons in our legal system,* while most living beings and living communities (bears, birds, whales, forests, coral reefs) have no rights as persons. Likewise, certain classes of human beings (blacks, women, Native Americans, Chinese) have been denied full rights as persons at times in our society when corporations were accorded full rights.

The economic and political power of Southern slaveowners caused the most egregious weakness and inconsistency in the United States Constitution—that which permitted the institution of slavery. The Fugitive Slave Act later ruled that escaped slaves—*even in states where slavery was prohibited*—had to be returned to their masters in slave states. Abolitionists opposed to slavery on moral grounds conscientiously broke this law by organizing the Underground Railroad to help slaves escape and then to move them through the United States to safety in Canada or secure areas in the North.

Lester Rhodes, writing in the *Earth First! Journal,* contends that this activity was an early example of monkeywrenching. Slaves were considered to be private property. By helping slaves to escape, abolitionists were destroying or stealing private property and at the same time extending rights to those excluded from the law. Similarly, ecoteurs destroy private property (e.g., bulldozers, survey lines) to extend rights to nature.

In *A Sand County Almanac,* Aldo Leopold writes, "The land ethic simply enlarges the boundaries of the community to include soils, waters, plants, and animals, or collectively: the land." Leopold discusses how in earlier human society, rights were not extended to women, slaves, and other classes of persons. They were considered property—which is what the land community today is considered

---

*At least corporations have the privileges of being legal persons; unfortunately they are not held accountable for their crimes as persons are.

to be. He argues for the extension of rights to the land and the creatures that compose it.

Dr. Roderick Nash develops in detail this historical progression of ethics in his book, *The Rights of Nature,* and compares groups like Earth First! and the Sea Shepherd Conservation Society to abolitionist groups before the Civil War.[5]

A political tactic. Monkeywrenching can also be seen as a sophisticated political tactic that dramatizes ecological issues and places them before the public when they otherwise would be ignored in the media, applies pressure to resource-extraction corporations and government agencies that otherwise are able to resist "legitimate" pressure from law-abiding conservation organizations, and broadens the spectrum of environmental activism so that lobbying by mainstream groups is not considered "extremist."

Unlike the other defenses of monkeywrenching, this is not a basic defense, except in a Machiavellian sense, but is a means-to-an-end defense that complements the other defenses.

There's nothing else left to do. I've previously mentioned the 1989 Hatfield rider on ancient forests, and the Mount Graham issue in Arizona. They are merely two examples of the failure of conservationists' "approved" methods to safeguard biological diversity. When the system leaves honest, committed defenders of nature no alternative, such people become monkeywrenchers.

A slightly different application of this approach is to see the value of monkeywrenching in saving important ecosystems until changes in society make such defense unnecessary. Ecotage can help save the building blocks of evolution that are now threatened and are unlikely to be protected by conventional means, until society evolves to an ethical level that holds wild nature to be important and worth preserving or until industrial society collapses and is no longer a threat to biological diversity.

5. Roderick Frazier Nash, *The Rights of Nature: A History of Environmental Ethics* (Madison: University of Wisconsin Press, 1989).

As a means of thwarting an unresponsive government.
Monkeywrenching is more than a collection of tools and techniques
to damage machines and inhibit development. It can also be used to
thwart the government and the industries that control it. America's
founding fathers were believers in John Locke's characterization of
government as an arrangement voluntarily entered into by free
individuals in a state of nature for the common good. Modern
historical and archaeological analysis presents a far different view of
the emergence of the state and the nature of government from that
articulated by Jefferson and Madison.[6] These studies lead one to
what I call the "Iron Law of Government": *The state, and all of its
constituent elements, exists primarily to defend, with the use of lethal
force if necessary, the power and status of the economic and philosophical
establishment.* This is not a statement of "ought"; it is simply one of
"is."

Thus we are left with the strategy of monkeywrenching: Don't
reform. Thwart. In addition to illegal ecotage, monkeywrenching
can be thought of as a strategy that includes entirely legal techniques
and that operates within the system, too—it is known as "paper
monkeywrenching" in such cases.

The most successful example of paper monkeywrenching was the
Wilderness Act of 1964. This landmark legislation was not a reform
measure, it was a monkeywrench in the gears of government. The
Forest Service proved itself unwilling to protect wilderness on the
National Forests when it began to hack apart Bob Marshall's system
and regulations after World War II. The Wilderness Act said simply that the Forest Service was incapable of protecting wilderness
values within its general framework of management, so decisions
pertaining to Wilderness would be taken out of its hands. The

6. The best analysis comes from Lewis Mumford. See particularly his two-
volume study of civilization, *The Myth of the Machine (Technics and Human Development* and *The Pentagon of Power* [New York: Harcourt Brace Jovanovich, 1964]).
Many books by Murray Bookchin also deal extensively with this topic. William H.
McNeill's *The Rise of the West* (Chicago: University of Chicago Press, 1963) is the
best comprehensive history of civilization.

Wilderness Act is an indictment of National Forest management. Designation of an area as Wilderness is a means not to reform but to thwart standard agency management. (Admittedly, some proponents of the Wilderness Act may not have realized they were monkeywrenching instead of reforming, but in analyzing the frustration that led conservationists to propose the Wilderness Act, we see that it is in fact a wrench in the gears.)

Similarly, the most important result of the National Environmental Policy Act has been that it has offered a handle for legal appeals of, and lawsuits against, agency decisions. It gives conservationists entry into a branch of government that is not part of the bureaucracy—the courts—and that can overrule agency decisions. Again, this is not a reform, it's a monkeywrench—albeit a legal, paper monkeywrench.

Of course, when paper monkeywrenches fail, there are more solid kinds to use.

George Washington Hayduke. There comes a time when the biological crisis is so severe, when the time is so short, when the forces of ecological destruction are so powerful, that the lover of the wild has no option but that taken by George Washington Hayduke in Abbey's *The Monkey Wrench Gang*. Bonnie Abbzug asks Hayduke how he proposes to bring about a counter-industrial revolution.

> *Hayduke thought about that question. He wished Doc were here. His own brain functioned like crankcase sludge on a winter day. Like grunge. Like Chairman Mao prose. Hayduke was a saboteur of much wrath but little brain. The jeep meanwhile sank deeper into Kaibab National Forest, into the late afternoon. Pine duff rose on dusty sunbeams, trees transpired, the hermit thrushes sang and over it all the sky (having no alternative) flourished its borrowed sundown colors— blue and gold.*
>
> *Hayduke thought. Finally the idea arrived. He said, "My job is to save the fucking wilderness. I don't know anything else worth saving."*

And so, ultimately, we come to a choice. Smokestacks, strip mines, clearcuts for crackerbox homes, DDT in penguin fat, topsoil eroding to the sea, whale meat on Japanese plates, spotted cat hides on vain rich humans, rhino-horn dagger handles on Yemeni belts, nearly four thousand million years of beauty, diversity, and life chewed up in the space of a human generation or two.

Or goose music, ancient conifer forests, spouting whales, rich complex tropical rain forests, elephants and rhinos and lions, coral reefs, howling wolves, clean air and clean water, nearly four thousand million years of beauty, diversity, and life sailing on through the blackness of space for another four billion years.

The Machine.

Or Life.

13 | Is Tree Spiking Necessary?

Let your life be a counter friction to stop the machine.

—Henry David Thoreau

Friday, May 8, 1987, 7:50 A.M. A band saw at the Louisiana-Pacific mill in Cloverdale, California, strikes an eleven-inch spike embedded in a redwood log. The saw shatters and pieces of blade fly across the room. One large section strikes off-bearer George Alexander, 23, in the face, breaking his jaw and knocking out several teeth.

Thursday, May 14, 1987. Louisiana-Pacific announces the incident to the media, blaming it on Earth First!, and offers a $20,000 reward for information leading to the arrest of the spiker. The media has a field day. At last there is an injury to lay at the door of "radical" environmental defense. Newspaper headlines scream "Environment Radicals Target of Probe Into Lumber Mill Accident," "SAWMILL: 'Eco-Terrorists' Are Focus of Probe," "Tree Sabotage Claims Its First Bloody Victim," "Earth First Blamed for Worker's Injury," "Tree-Spiking 'Terrorism' Blamed for Injuries." . . .

Columnists and editorialists unleash their wrath: "Evils of Spiking," "Anarchy Is Busting Out All Over the Lot," " 'New Terrorists' Must Be Fought," "The 'Green Bigots' Turn Bloody," "Eco-fanatics Endanger Workers' Lives."

Local politicians in northern California jump on the Louisiana-Pacific bandwagon to denounce "violent" environmentalists, and Idaho senator James McClure gravely takes to the Senate floor to pronounce the incident "a gross, premeditated attempt to injure or perhaps even kill someone to draw attention to an environmental cause."

The impression is created by the timber industry, pro-logging politicians, and the unquestioning media that Earth First! had spiked a giant redwood in the hopes of injuring a logger. Other environmentalists and even a few Earth First! supporters respond by denouncing monkeywrenching and calling on all to work within the system.

It was enough to make even George Washington Hayduke stop and reconsider.

In the years since Anderson's tragic accident, this impression has been used to stir anti-wilderness sentiment into a frenzy by timber companies and politicians. Montana congressman Ron Marlenee encouraged timber workers at a 1989 election rally to "spike an Earth First!er," and crowed that when the good ol' boys got done with them, they'd look like "trampled horse dung." McClure successfully used the incident to attach a rider to the omnibus drug bill in fall of 1988 that made tree spiking a federal felony, with draconian penalties for those convicted. In magazine and newspaper articles about Earth First!, even those that are generally evenhanded, the Cloverdale incident has been repeatedly dredged up. Tree spiking has been successfully labeled as a violent, life-threatening tactic, and it has been effectively linked to Earth First!; indeed, Earth First! has become almost synonymous with tree spiking.

I admit that I have come to question the utility of tree spiking because of the Cloverdale incident and its aftermath. It is time for advocates of monkeywrenching to reconsider this tactic. We need to determine if tree spiking is effective, safe, and ethical.

Before we can make that determination, however, we need to delineate clearly what tree spiking actually is, and define its purpose, and we need to bring out the facts of the Cloverdale incident. Let us do the latter first.

The timber industry and anti-conservation politicians characterize the Cloverdale incident as follows:

■ It was done by Earth First! or at least by other radical environmentalists.

■ It was done to protect old-growth redwoods in a controversial sale area.

■ It was done without any warning or regard for the safety of timber workers.

■ It is typical of other tree spiking or monkeywrenching incidents.

■ It is typical of what will generally happen if a spiked log is run through a sawmill.

These are the impressions that linger after Alexander's injury; these five assumptions are now connected with tree spiking, monkeywrenching of any kind, and Earth First! in general. But are they correct?

■ After the initial hoopla blaming Earth First! for the accident, several northern California newspapers issued apologies when it was learned that the Mendocino County Sheriff's Department's primary suspect was a conservative Republican in his mid-fifties who owned property adjacent to the logging site (although he was believed to live in southern California). Absolutely no evidence has ever been presented connecting an Earth First!er or other conservationist with the spiking.

■ The spiked tree was not an old-growth redwood from a sale in a pristine area. It was a seventeen-inch-diameter, second-growth redwood from a decidedly non-wilderness tract. (For perspective, a telephone pole is about seventeen inches in diameter.) Conservation groups had not been opposing this particular sale, although area residents were upset by heavy truck traffic, noise, and erosion resulting from the cutting.

■ Although Louisiana-Pacific claimed they had received no warning, they later admitted that dead animals had been previously left on machinery in bizarre protests against the sale. Other evidence indicates that Louisiana-Pacific may have known some of the logs from the sale were spiked. In fact, the day after Alexander's injury, L-P ran another log from the same sale through the band saw and hit a similar spike. Fortunately, no one else was injured. Until after the second spike was hit, the company excused itself from using metal detectors because of expense and inefficiency.

■ No other injury is known to have resulted from tree spiking in the United States, Canada, or Australia (the countries where spiking is widespread). Since tree spiking is meant to save trees, trees are properly spiked when they are standing and alive. The tree that injured Alexander had been spiked after it was cut down and bucked up. The spike was driven into the butt end of the log and counter-sunk by another spike. By freak chance, the band saw hit the spike head on—in other words, it didn't have to simply cut across a three-eighths-inch-diameter nail, it ran into the full length of an eleven-inch nail.

■ After the initial sensationalism about the incident, careful reporting revealed that the accident was more of a mill-safety issue than one of dangerous tree spiking. In a copyrighted interview with Alexander in the *San Francisco Examiner*, reporter Eric Brazil quoted the injured man as saying the band saw was cracked, wobbly, and due for replacement. Alexander said that he had been complaining about the dangerous condition of the saw for two weeks. He also said that the saw hit metal about "four times a week." "If it had been a good saw, it would've handled the spike better," Alexander said. He further stated that he almost had not gone to work on the fateful day because of concerns about the saw.

Tree spiking is by no means a recent invention. It was used around the turn of the century by loggers during the labor wars with the big timber companies in the Pacific Northwest. Earth First! initially learned of it in the early 1980s from loggers in Montana who were opposed to the cutting of ancient forests by large corporations.

My book *Ecodefense* treats the practice in detail. Quite simply, tree spiking consists of inserting a large nail or other piece of metal, or a hard, nonmetallic object like rock or ceramic, into a tree. Timber companies bidding on the sale, the Forest Service, or other agencies are then warned. If the Forest Service decides to sell the timber anyway, they have to send a crew into the woods with metal detectors and crowbars to remove the spikes. In several cases, this has cost more than the sale was worth. Many lumber mills have installed, at additional expense, their own metal detectors. Spiking a tree at a low level, where the chain saw used in felling the tree may encounter the spike, is discouraged for safety reasons.

There are techniques available to the tree spiker that make it very difficult to find or remove spikes: snipping off the heads of the spikes; using helix spikes; using nonmetallic spikes, which do not activate metal detectors; and using spurs and lanyard to climb high into the trees to spike, thereby preventing detection in the forest (although the spike can still be detected in the mill).

If a spiked tree is cut down and run through a lumber mill, the spike can damage the blade of the band saw to the tune of a thousand dollars or more. If the head rig is damaged, repairs can cost $20,000. It is only worn-out, unsafe saws such as the one at Cloverdale that can shatter on impact.

With these facts, advocates of monkeywrenching can reconsider tree spiking and determine whether it is safe, ethical, and effective. Unless it meets all three criteria, I believe it can no longer be justified, and that ecodefenders should discontinue its use.

Is tree spiking safe?

Safety is a relative concept; nothing in life is entirely safe. The record shows, however, that tree spiking is one of the least of dangers a logger faces.

Timber work is one of the most dangerous occupations in North America. The *Fallers' and Buckers' Handbook* of the Workers Compensation Board of British Columbia reports that in the period 1971–80 there were 175 fatalities, an average of seventeen per year, from field logging. Mill accidents also killed many people. In the state of Washington during 1988 there were twenty-eight fatalities in logging accidents. Statistics indicate that one logger in six will have his career ended by a fatal or crippling injury. Logging accidents are currently increasing in number because big timber companies busted unions in the 1980s, thereby removing or weakening union oversight of worker safety; more large timber companies are utilizing generally nonunion logging contractors who do not place a priority on safety; and companies borrow money at high interest to buy logging equipment, then maximize production for a quick return to pay back the loan, thereby causing rushed, corner-cutting working conditions.

These figures compare with one injury sustained by tree spiking—and that from an irresponsible, atypical operation. The danger to woods and mill workers comes not from tree spikers trying to defend old growth, but from large companies who rush poorly protected workers in the interest of profit for their stockholders.

Responsible tree spiking always involves warning potential logging contractors and the managing agency that a sale has been spiked. This can be done through a communiqué or phone call, by spray painting a large *S* or similar warning on spiked trees or on the edge of the sale area, by flagging a spike high in a tree to warn of high, unflagged spikes in other trees in the stand, or through other unmistakable means.

If such a warning is not heeded by the Forest Service or a mill operator, it becomes a matter of debate as to who is responsible if a saw encounters a spike. In all cases where this has occurred (and there have been several), in only one instance—Cloverdale—did an injury result. History suggests that if a band saw is in proper operating condition, hitting a spike will not cause injuries. Metal also finds other ways of ending up in saw logs: barbed wire from old fences, nails from hunting camps, nails used to attach insulators to trees for ancient Forest Service phone lines, and so forth. Such metal commonly causes damage to band saws in sawmills—much more frequently than do spikes placed by monkeywrenchers. Trees along logging roads are generally so full of target-practice bullets that they cannot be milled.

There is a thoroughly safe solution to the problem of spiked trees: Do not cut them down. For timber operators to be able to make that decision, though, they must be given a timely and insistent warning that the trees in a particular sale have been spiked. In one incident, on the Prescott National Forest in Arizona in 1989, the Forest Service was warned about a spiked timber sale, but neglected to tell the timber operator because it was "too much trouble" to check the trees to see if they had actually been spiked. In this case, a spiked tree did $900 worth of damage to a band saw, and the timber company considered suing the Forest Service. No one was injured. A similar problem involves the frequent selling and trading of logs from one mill to another. An unethical mill might resell logs they know to be spiked to another mill and pretend not to know they were spiked.

Is tree spiking ethical?

This is a question open to great dispute. Those who believe in the sanctity of private property, who believe that laws must never be broken, or who believe that forests are merely resources to be used by people will always argue that tree spiking is immoral.

I maintain that tree spiking can be ethical, for several reasons.

Forests are not simply collections of trees. All natural forest eco-systems (including seral forests, born in natural wildfire, and recovering second-growth forests in the East) are integrated, complex systems. Old-growth forests, the most developed and complex forest ecosystems, cannot be re-created or regrown within human time frames. Many species depend on natural forest ecosystems. These species and the community they form have value in and of themselves, not merely for the benefit of human beings. They are products of the same process of evolution that produced us, and have intrinsic value.

Current logging of old growth on Forest Service, BLM, and private lands represents the final mopping-up in a two-century-long campaign of genocide against the ancient forests that blanketed much of the United States. Only 4 percent of the Coast Redwood forest remains intact; half of that is under immediate threat through the activities of Maxxam Corporation. Scarcely more of the old-growth conifer forests of Oregon, Washington, and California remain. They are scheduled for liquidation within the next several decades. Elsewhere in the Western United States, and in British Columbia, Australia, and southeast Alaska, ancient forests are being similarly devastated. In the Eastern United States, the Forest Service has begun to road and clearcut second-growth forests that have recovered mature characteristics after decades of protection. Defending the victims of this genocidal campaign seems to me inherently ethical.

Efforts by conservationists to preserve viable old-growth ecosystems by working within the system are failing. Wilderness legislation has preserved little of the ancient forest; in most cases, the lush old-growth forests at lower elevations have been left out of designated Wilderness Areas so that they can be logged. Hikers get the scenic "rocks and ice" Wilderness loggers get the trees, and biodiversity gets the shaft. When conservation groups file lawsuits to halt timber sales based on federal statutes such as the Endangered Species Act, the National Environmental Policy Act, or the National Forest Management Act (NFMA), members of Congress in thrall to the timber industry quickly pass legislation voiding or preventing such suits. Protesters using nonviolent civil disobedience to slow logging of wild forests are being hit by lawsuits from timber contractors. One group of Oregon protesters, the Sapphire Six, served two weeks in jail and was ordered to pay the logging contractor $35,000 for downtime. This leaves monkeywrenching.

Therefore, responsible tree spiking (done as a last resort after legal means, civil disobedience, and lesser forms of monkeywrenching have failed; and only with full warning to the land-managing agency and timber harvesters) is justified and ethical.

But is tree spiking effective? Does it work? Can it save trees? Here we step out of our canoe onto trembling ground. Two arguments against the effectiveness of tree spiking deserve careful, even agonized, scrutiny.

The first is that tree spiking gives the anti-wilderness voices ammunition to turn the media and the public against conservationists and protection of our forests.

Despite the logical explanation of the facts of the Cloverdale incident and why tree spiking is necessary, pandering politicians will continue to use the issue of spiking to whip up sentiment against "environmeddlers," timber companies will continue to use it to elicit public sympathy, and the media will continue to misrepresent it for better copy. No matter how careful or how responsible Earth defenders are in tree spiking and in publicly defending the practice, tree spiking will be held up as proof that monkeywrenchers are more interested in trees than in workers' lives, that ecoteurs are "ecoterrorists."

We must keep in mind, though, that if tree spiking did not exist, the same calumny directed against it would be directed against decommissioning heavy equipment;[1] if that did not exist, it would be against survey-stake pulling; if that did not exist, it would be against civil disobedience; if that did not exist, it would be against Sierra Club lawsuits and Wilderness proposals, just as it was in the 1970s (and still is in some places). I recall rhetoric and threats from the timber industry, their suckered workers, and their captive politicians during RARE II against the Sierra Club and local conservationists that were just as inflammatory and jingoistic as what is used today against tree spikers.

The second argument is that tree spiking simply has not worked. The Forest Service, with metal detectors and crowbars, has gone into sales that have been spiked and removed the spikes. Although

---

1. A conservationist from the Northwest reports that damage to equipment actually causes more anger in logging circles than does spiking—even though there is no human safety issue involved. He speculates that spiking gets more response only because the media reports it more, since spiking is more interesting and unusual than mere "vandalism." Might this indicate that loggers really do not feel that spiking is as dangerous as they publicly claim, or does it indicate that decommissioning heavy equipment is more effective at stopping logging than is spiking?

it may cost more to make the sale safe for cutting than the sale is worth, the Forest Service does it to save face and to prove that the agency will not be "held hostage" by monkeywrenchers. (Senator McClure even goes so far as to threaten *designated* Wilderness Areas with clearcutting if tree spiking does not stop. He is considering legislation, or so he says, that would mandate the logging of one hundred acres of designated Wilderness for every acre of timber sales spiked. Of course, it is extremely doubtful that such a childish ploy could pass Congress. One would expect it to incite another national outcry such as the one that occurred when Jim Watt tried to offer oil-drilling leases in designated Wilderness Areas.)

There are two counterarguments to this assertion. One is that, in fact, tree spiking has stopped several timber sales in the states of Washington, Oregon, Virginia, Colorado, New Mexico, and Montana. On British Columbia's Meares Island, a massive and carefully organized spiking campaign played a role in stopping the clearcutting of the giant cedars there. It would be helpful to survey other spiked sale areas to determine how many have been quietly spared from chain-saw massacre. (Forest Service investigators will not release their figures on the effectiveness of tree spiking, so as not to encourage more spiking.) The other counterargument is that the timber industry, politicians, and log-town editorialists would not be weeping and wailing about tree spiking if it were not indeed effective at stopping logging.

But is tree spiking really effective? Is it of significant value in stopping the logging of our forests? Probably. In some cases. But . . . I dunno. It's like a tough piece of jerky being chewed around the campfire. You chew and you chew and you chew and nothing much happens. You work up a lot of spit, but you still have a big glob in your mouth.

I dunno.

If we accept for the sake of argument, however, that tree spiking can be effective in halting logging, that it is a tool that needs to be retained in the monkeywrencher's kit, we should take steps to defuse the political tension surrounding it.

First, ecodefenders should recognize spiking as an extreme measure, to be used only when other means fail. However, the safest time to spike a potential timber sale is years before it is to be cut. There is less likelihood of encountering forest rangers or timber bidders in the sale area then, and spiking years in advance allows the tree to grow bark over the spike, disguising it and making its discovery and removal more difficult. Therefore, some might argue that prophylactic spiking well in advance is the best method. If conventional means later stop the timber sale, fine; if they don't, then the spikes are already in the trees and a warning can be made. When monkeywrenching is necessary, less sensational methods such as desurveying, road removal, and heavy-equipment decommissioning should be utilized in lieu of tree spiking. Regularly damaging key sections of the unpaved road network and new road construction on the National Forests might do more to halt ancient forest logging than spiking does.

Second, tree spikers should be absolutely rigorous in minimizing the potential for injury in spiking. Unmistakable, direct warnings (not threats!) must be given to companies planning to cut or mill timber from sales that have been spiked. Such warnings should be given with humility, not with taunts and dares.

Third, publicity about tree spiking should be muted. Warnings to the Forest Service and logging companies are necessary, but notice to the media should be discouraged. This allows the timber beasts the opportunity to drop the sale quietly and not lose face, and it doesn't alienate community support. The media should be notified about a spiked sale, however, if it appears that the Forest Service or logging contractors are going ahead with the sale anyway, figuring that if an injury occurs from milling a spiked log without checking it first, good (from their point of view) publicity will result.

Fourth, monkeywrenchers should consider another potential problem with tree spiking. Long after a spiked timber sale is canceled, an innocent firewood cutter might hit a spike with his chain saw. Of course, firewood cutting is possible only where there are roads, so one solution would be to spike only in roadless areas and to warn the Forest Service and potential contractors before roads are built into the area. This may not be a major problem, but it should be taken into account before a stand of trees is spiked.

Finally, forest defenders must articulate the need for preservation of our remaining ancient forests and other natural forests (including recovering second growth) as effectively as possible. We must continually try to get our message across to the public in an understandable way.

Tree spiking may be an extreme tactic, but it may also be a necessary one. The United States, Canada, and Australia have devastated their ancient forests to a greater extent than have most tropical nations their rain forests. The rapacious forestry practices causing the extinction crisis are every bit as rampant in our "developed" countries as in the Third World. A tiny fraction of the ancient temperate forest remains—perhaps too little to sustain the complex communities that have evolved over millennia. The remnants must be protected. If asked, the Northern Flying Squirrel, Western Red-backed Vole, Marbled Murrelet, Northern Spotted Owl, Pine Marten, Black-tailed Deer, Pacific Giant Salamander, mycorrhizal fungi, Douglas-fir, Western Hemlock, Sitka Spruce, Western Redcedar . . . would agree. There can be no compromise.

14 | Strategic Monkeywrenching

This is the real world, muchachos, and you are in it.

—B. Traven

The Doberman down the block starts barking its brains out at first light. I put in my ear plugs, roll over, and go back to sleep. A couple of hours later, I hear the sound of many feet in the hall outside my bedroom, and an unfamiliar but authoritative voice yelling my name. Disoriented from the sudden awakening, unable to hear clearly through the ear plugs, I open my eyes to three men around my bed, pointing .357 Magnums at me.

I wonder for a moment where Alan Funt is—but no, this isn't Candid Camera. These guys really are FBI agents, and I am under arrest. . . .

The human brain is constructed so that it requires a swift kick several feet beneath it before it comprehends new realities. So it was for me, and, I think, others of the ecodefender persuasion in appreciating that the government and corporate reaction to monkeywrenching had changed in the last two years.

Not that we didn't have fair warning that the advocacy and practice of monkeywrenching was entering a more difficult era, mind you. The Louisiana-Pacific sawmill accident in 1987 elicited bug-eyed cries of "eco-terrorism" from the hacks and pimps for fast-buck rape-and-scrape industries. The attachment of Senator James McClure's anti-tree-spiking rider to the omnibus drug bill in late 1988 should have warned us that the spooks were pulling on their cloaks and sharpening their daggers. But it is only now, as the full scope of the FBI blitzkrieg against Earth First! becomes obvious, that we understand ecotage has passed through a transition period. Its use as a tactic, strategy, and symbol for defense of natural diversity now confronts a new, more dangerous environment.

Reading FBI documents about their $2-million campaign to frame me and other activists, reading the history of the COINTEL-PRO operations to destroy dissident groups twenty years ago, it begins to sink through my thick Neanderthal skull that if monkey-wrenching is to remain an effective tool for Earth defenders, its practitioners must adapt to the new situation. If they fail to adapt, they will quickly become extinct as the power structure comes down upon them.

Juvenile bravado and hotheaded posturing aren't useful in facing this challenge. Courageous resistance to the destruction of the planet continues to be essential, but I doubt that tragic heroes or martyrs are the roles that must be played. Notwithstanding the blithe-spirited "radicalism" of some, intelligence, cleverness, deliberateness, and maturity are not fundamentally inconsistent with the no-compromise defense of Earth. Indeed, these qualities are absolutely necessary. Monkeywrenching is a direct outgrowth of guerrilla war theory, and the successful guerrilla is the one who returns to fight again and again.

For monkeywrenchers to adapt to new conditions, they need to be open, creative, and *smarter*. While I would not encourage anyone to monkeywrench (that is an entirely personal decision),[1] and while

1. More important, I would not want to *dis*courage anyone from monkey-wrenching. Those willing to commit ecotage are needed today as never before. The advice I offer here is merely that—advice. I respect the arguments of those brave souls who proclaim that it is necessary to take risks in defense of natural diversity, that the battle is far more important than our individual liberty and safety. My purpose is to propose methods for monkeywrenchers *who want to minimize the*

I have of necessity hung up my pearl-handled wrenches for good, if I were sitting around a wilderness campfire with a few old friends, smoking a good cigar, and musing about the future of monkey-wrenching in the most general of terms, I might say to George and Bonnie that I have a few ideas for monkeywrenchers who don't want to get caught and who want to be as effective as possible.

Monkeywrench alone or with few and absolutely trusted partners. While a partner or partners can increase a monkeywrencher's effectiveness and enable her or him to take on bigger and more heavily guarded targets, and sometimes practice better security, most monkeywrenching can be safely done by one person. Moreover, if they operate alone, ecodefenders need not worry about partners with loose lips, infiltration by informers or agents provocateurs, or betrayal by weak-kneed compatriots trying to save their own skins. If one chooses to practice ecotage with partners, however (and we must recognize that as social critters, we generally want to do things with friends), there should be no doubt as to their trustworthiness. *If* they work with others, mature monkeywrenchers work only with those to whom they would entrust their lives, for that is what is being entrusted. The number of people involved should be the absolute minimum necessary.

Keep a low profile. Unfortunately, being a visibly active Earth First!er, or even attending Earth First! events, makes one a suspect for law-enforcement agencies tracking down monkeywrenchers. Infiltrators will target EF! gatherings and groups to set up potential ecoteurs. Bored cops pretending to be James Bond take telephoto portraits of participants at Earth First! rallies, or write down license plate numbers of vehicles outside such events. The most effective monkeywrenchers will be those who keep invisible, who seem to mind their own business, who appear to be apathetic and uninvolved in causes.

risks, and to suggest ways to make monkeywrenching more palatable to the general public and less likely to elicit accusations of "terrorism."

Given the fervent interest of the gendarmes in busting every monkeywrencher, it may not even be safe for an active monkeywrencher to be visible within mainstream groups like the Sierra Club.

Avoid sporting "Hayduke Lives" bumper stickers or patches, or other visible emblems. Any conservationist bumper sticker may mark one as a suspect in some rural areas. Careful ecoteurs may even try camouflage—an American flag decal or NRA sticker. Non-monkeywrenchers like me should continue to brandish "I'd Rather Be Monkeywrenching" bumper stickers.

Keep *Ecodefense* out of sight. Possession of this book will be considered tantamount to an admission of guilt. It shouldn't be displayed in ecodefenders' vehicles, homes, or places of work. They shouldn't show it off. They may read it, study it, and use it, but they should not let it draw suspicion to them.

Keep a closed mouth. Ecoteurs should talk to *no one* about their exploits. If they do brag to anyone, they should at least avoid mentioning anyone else with whom they have monkeywrenched. *They should never incriminate anyone under any circumstances!*

A little-appreciated danger of telling others (particularly trusted friends) about one's exploits is that they might be called before a grand jury. *If* they are offered immunity from prosecution, they cannot invoke their Fifth Amendment rights, and can be forced to testify about what they know about other people. If they refuse, they can be jailed. If they lie to save a friend, they can be convicted of perjury and imprisoned. Monkeywrenchers should never put a friend in the position of having to go to jail to keep from incriminating them.

Do not draw media attention. Previously, judicious publicizing of monkeywrenching was useful in making others aware of the widespread nature of the practice and the need for it. This encouraged potential monkeywrenchers and intimidated wilderness destroyers. Monkeywrenching has now received adequate media attention, it can be argued, and can now be carried on effectively without publicity. Warnings about tree spiking should still be made to the Forest Service or logging contractors, of course. And, in certain cases, publicity about an action may still serve a worthwhile purpose; this must be determined individually.

Avoid targeting national security-related targets or major industrial facilities. Activities against such sites draw in the FBI or other specialized units to investigate, court more serious penalties, elicit greater public and media condemnation, require greater security and planning, often necessitate a larger group of ecoteurs, and historically have accomplished little.

Avoid those who talk about monkeywrenching. If someone begins to talk about specific monkeywrenching, walk away or hang up.

Avoid or ostracize lunatics, advocates of violence, or immature, macho bigmouths. These are the people who will get monkeywrenchers into trouble personally and discredit environmental groups, whether they are infiltrators, crazies, or merely fools. Exclude them from your own life and from environmental events.

Minimize the use of monkeywrenching techniques that might be characterized as "violent." Violence here should be defined broadly. Careful monkeywrenchers avoid explosives, firearms, and arson, and seek alternatives to tree spiking whenever possible. They diligently practice safety measures and conscientiously concern themselves with the welfare of those who may come into contact with their operations.

Avoid public statements or writings that can be construed as advocating violence. Again, ecodefenders should define "violence" broadly. On a related topic, they ought to avoid making comments such as "they'll never take me alive." Such braggadocio may cause ecoteurs to be denied bail after being arrested.

Be willing to disavow stupid acts. Actions such as the arson of the Dixon, California, livestock auction barn, the Earth Day 90 sabotage of power lines in the Santa Cruz, California, area and the burning of an American flag at the 1989 Earth First! Round River Rendezvous, whether committed by fools or by persons deliberately trying to discredit Earth First! (loggers, law-enforcement agents, etc.), will increase. Spokespersons for Earth First! and other conservation groups must be careful in commenting on dubious events. They should not condemn a well-done and properly targeted act of ecotage, yet they should not let themselves be conned into supporting a stupid one. Often it is best to decline to comment if one possesses incomplete information about the circumstances.

Avoid the use or possession of illegal drugs. The use of illegal drugs presents several problems for monkeywrenchers. If they are arrested for monkeywrenching, drugs add to the likelihood of their being denied bond, being convicted, and receiving prison sentences. Being involved both with monkeywrenching and drugs doubles an individual's chances of being arrested. The federal government is engaged in a massive crackdown against drugs (it can be argued that the purpose of the current antidrug campaign is to create the public support and apparatus for a more authoritarian state), and the use of drugs may place one under surveillance that may uncover ecotage activities. If monkeywrenchers are serious warriors for Earth, they will minimize things that may draw attention to themselves or jeopardize their operations.

Of course, the above comments are just maunderings around the fire, smoke rings blown into the night air, desultory accompaniments to the hooting of owls. The kind of casual talk you might have heard in a seedy Boston waterfront tavern in, say, 1773. . . .

15 | The Perils of Illegal Action

Certainly one of the highest duties of the citizen is a scrupulous obedience to the laws of the nation. But it is not the highest duty.

—Thomas Jefferson

S ome of the perils of conscientiously disobeying the law quickly become apparent to anyone who chooses to do so. The indignity and boredom of arrest, booking, incarceration, and court proceedings can be nearly insufferable. Sparring with the legal system costs money, time, and energy. Finally comes the penalty, with further loss of money (fines) or freedom (jail sentences). Other hazards may arise as well. The Sapphire Six, who occupied a logging site in Oregon, have been sued by the contractor for downtime. Texas Earth First!er James Jackson injured his leg when a Forest Service officer chopped down the tree in which he was sitting. Peace activist Brian Wilson lost his legs to a train. Students who campaigned against tyranny in Beijing have been lined up against the wall. When one engages in deliberate civil disobedience, one quickly begins to understand Mao's maxim, "Political power grows from the barrel of a gun."

Having just been arrested while asleep in my bed by a posse of gun-wielding FBI agents playing Dirty Harry, and now facing a possible five-year sentence in a federal pen on a set-up charge, I have no desire to downplay these dangers. Anyone who chooses to stand against a corrupt and brutal establishment (and, to varying degrees, all political states are such) must accept that he or she may eventually face that ultimately lonely moment shared by Joan of Arc, Nathan Hale, and the French revolutionary Georges-Jacques Danton.

But there are other kinds of pitfalls in choosing to break the law, more subtle than those above, but no less dangerous.

One danger is that by conscientiously breaking unjust laws or by carefully targeting wilderness-destroying property for destruction, one places oneself in opposition to the creators, beneficiaries, and enforcers of those laws, or to the owners and users of that property. It is an easy step from that to creating a dualistic world of *Us* versus *Them*. When we create such a world, our opponents become the enemy, become the *other*, become *evil men and women* instead of men and women *who commit evil*. In such a dichotomous world, they lose their humanness and we lose any compulsion to behave ethically or with consideration toward them. In this psychological state, we become "true believers," and any action against the enemy seems justified. One needs only to look at Adolf Hitler or the Ayatollah Khomeni to see the damage to one's psyche that results from holding such attitudes.

Another peril of the lawbreaking process is a loss of focus. For practitioners of civil disobedience or ecotage in defense of natural diversity, the fundamental issues are wilderness and wildlife. Our opponents are federal land-managing agencies and resource-extraction industries. After arrest, incarceration, and court sentencing, however, it is easy to become confused and begin to see the injustice of the legal system as a fundamental issue with which we must deal, and to begin to regard the deputies, jailers, and judges whom we encounter as our primary opponents. When this occurs, our focus on wilderness is diluted. It is important to preserve that focus.

The adrenaline rush of a well-planned action may seem an effective counter to the dull security and safety provided by modern society. Some turn to thrill sports like rock climbing, skydiving, or dirt-bike racing for the same rush. The monkeywrencher may become captivated by the intoxication of destroying machines and getting away with it. The tingle of action may be a justifiable part of the reward for courageous defense of wildness; it becomes a danger when it turns into a delirium or is the primary reason for breaking the law.

Often a key element in civil disobedience or monkeywrenching is gaining public acceptance or understanding of the injustice of certain laws. If our ethical disobedience becomes unfocused, untargeted, and ethically ambiguous, then we appear to the public as hooligans and common criminals. If lawbreakers for a good cause do not act deliberately, then the ethical statement they make is demeaned, and it is easier for those in power to turn the public against just causes.

The greatest hidden peril of illegal action is simply that when one breaks the law, even an unjust law, with regularity, breaking the law can become seductively easy. It becomes common, even normal, to break the law.

Although the laws of a modern state are created by and for an economic elite to maintain their financial position and to defend the philosophical orthodoxy to which they subscribe, many laws are nonetheless necessary when millions of people live in close proximity. All human societies have customs and rules governing interactions between and among individuals. They are natural; they should be obeyed. I believe in laws against rape, assault, and invading Wilderness Areas with vehicles or chain saws.

The more one becomes involved in conscious lawbreaking, whether nonviolent civil disobedience or monkeywrenching,[1] the more one needs to be scrupulously deliberate about doing so. Without such fastidiousness, one risks damaging one's own psyche and

---

1. I should acknowledge here that public civil disobedience and covert monkeywrenching are generally considered entirely separate strategies, and that very different people engage in them. Although both involve consciously breaking the law, for many monkeywrenchers, breaking the law is incidental. Their aim is to thwart destructive machinery threatening natural diversity. Such tampering with machinery, however, happens to be illegal. As such, monkeywrenching shares the perils of civil disobedience discussed here.

one's cause. When we break unjust political laws to obey higher
ethical laws, we must guard against developing a laxity toward
standards in general. Indeed, when one deliberately engages in civil
disobedience from time to time, one needs to attend to *just* laws with
an even greater sense of responsibility.

Some who are deeply committed to the defense of Earth and to
opposing tyranny would undoubtedly disagree with the above.
Some people who have engaged in ecodefense actions may argue
that they have no obligation to honor any of the rules or customs
of this society, that they are free agents, or that they are in the
process of creating a new society with a new morality.

I wish I were so sure of myself. It would be an easier, simpler
world. It was so, for the heroes of matinee Westerns when I was
growing up in the fifties. I wanted to be like them—strong, silent,
secure, and whole in myself.

But I find that I cannot stand apart from or above society in that
way. How do you change society when you are apart from it? How
do you understand yourself when you deny the social environment
that produced you? How can you gain support for your goals and
actions when your behavior alienates potential supporters?

Wise guerrillas know that they are part of society and need support from the population base. The isolated, alienated guerrilla is
just as lost and vulnerable as the isolated, alienated Gorilla. We
primates are social animals. We have a long, deep heritage of being
part of a tribe, of defining ourselves by the cultural context in which
we were born.

We deny human ecology when we argue that we can operate
totally apart from the mores of society, when we define ourselves
as ethical islands, beholden to no one, without responsibility to
others for our own actions. There we enter uncharted waters,
beyond anthropology, beyond biology, into modernist alienation
and nihilism, into Hobbes's nightmare of all against all, a dark and
fearful place as far from the wilderness as we can imagine.

16 | Some Thoughts on True Believers, Intolerance, Diversity ... and Ed Abbey

Be as I am . . . a part-time crusader, a half-hearted zealot.

—Edward Abbey

There are more goblins lurking alongside the political activist's trail than I can number. One of them is the possibility of becoming a True Believer. It may be the biggest and most dangerous.

We all know the characteristics of the True Believer: He loses his sense of perspective and believes that his issue is the most important one; he often becomes egotistical, and thereby excessively critical of and impatient with others; he may fall into patterns of overwork where he believes he is indispensable, that no one else can do the job.

The deadliest trait of the True Believer, though, is a loss of tolerance for other approaches, for anyone whose ideas are not "politically correct." Political correctness is determined not just by the issue at hand, but by a whole litany of issues and positions. Those who don't fall into line on every point are completely out of line. The True Believer believes he has the *only* valid way. Those not on the True Believer's path are either foolish dupes or conscious agents of evil.

This is why the most vicious cat fights occur *within* a particular cause: True Believers most often hiss and spit and claw at those who display minor heresies from the revealed truth. Such self-defeating madness is repeated over and over in the history of civilization.

The self-righteous surety of True Believers notwithstanding, diversity in social and political causes is as important as it is in ecosystems. (An ecosystem with only one predator of large herbivores, for example, is generally less stable than one with several such predators. An animal that depends exclusively on a single species of plant is in a more precarious position than one that is able to feed off many different species.) For rising to the extraordinary challenge posed by human destruction of biodiversity and by the concomitant domination and exploitation of people by an elite and brutal oligarchy, diversity is particularly essential. The problems facing us are so vast, so multifaceted, that there is no one true path, no perfect answer. We need many paths; we need to ask many questions. Numerous styles are available and appropriate; there are countless tools suitable for tackling different aspects of each problem.

We did not form Earth First! with the thought that we had the only proper methods: confrontational civil disobedience, monkey-wrenching, and uncompromising advocacy. We founded Earth First! because these particular tools were not otherwise being used in defense of native diversity. Just as the Earth First! approach to the global ecological crisis is important, so are those of The Nature Conservancy, Sierra Club, Audubon Society, The Wilderness Society, Defenders of Wildlife, Earth Island Institute, Rainforest Action Network, Cascade Holistic Economic Consultants, Society for Conservation Biology, Negative Population Growth, Natural Resources Defense Council, Citizens Clearinghouse for Hazardous Waste, Worldwatch Institute, Greenpeace, Cultural Survival, bioregionalists, Green parties, and countless local groups. I may have suggestions for making these other approaches more effective, but purchasing land for conservation purposes, lobbying Congress and agencies on ecological issues, filing environmental appeals and lawsuits, conducting scientific and economic research into the value of wild nature, and developing alternative soft-path life-styles are all valid and necessary methods, just as are the hard-ass, court-of-last-resort avenues of the Sea Shepherd Conservation Society and Earth First!

We need more diversity, not less, in the effort to protect three and a half billion years of evolution from plundering by our international industrial growth society. Every available tool needs to be employed; every style, from business suits and laptops to camouflage and tree spikes, needs to be encouraged. The conservationist spectrum needs to be fleshed out and thoroughly filled. More questions need to be asked; more provocative, original answers offered.

As a practical matter, however, diversity under one roof has its limits—which is one reason why many independent groups are needed. Consensual decision-making works only when there is general agreement on worldview and proper approach. If a group is too diverse, then too much time is wasted debating strategy and philosophy, and too little real work is done. For example, even though The Nature Conservancy and Earth First! have similar goals of saving native diversity, our techniques are so different, our styles so divergent, that the two could never federate into a single organization!

In some cases, as a group grows, it becomes more and more diverse, spreading its umbrella farther and farther afield to include more styles and tactics. When this diversity becomes counterproductive—when there is no longer general agreement on philosophy, style, strategy, and tactics, when there is continual bickering between different factions, when internal dissension prevents the real work from being done—then it is time for a no-fault divorce. The different elements need to go their separate ways, without recriminations, without anger, each recognizing the worth of the other's position, but realizing that real differences exist. As a specific example, in the 1984 civil-disobedience campaign to save old-growth forest in Oregon, there was an excessive gap between adherents of philosophical nonviolence and rowdy Earth First!ers. The Gandhians objected to the clenched fist of the Earth First! logo and to calling Forest Service officers "Freddies" as being aggressive. Earth First!ers felt that informing law-enforcement agencies beforehand about planned actions was counterproductive because it would lessen the likelihood of achieving a shutdown of logging operations.

The Cathedral Forest Action Group separated and did effective work. If we had all tried to stay together under the Earth First! banner, energy would have been wasted in endless discussions between people not sharing fundamentals.

But what does any of this have to do with Ed Abbey?

Edward Abbey was not a True Believer.

Ol' Cactus Ed had strong opinions, yes. He was wildly in love with the wild. He believed in defending what he loved, and in defending it to the hilt. He loved a good argument, and he loved deflating pompous True Believers.

But Ed did not take himself too seriously. He poked fun at himself as often as he poked it at others. He created a caricature of himself. And the humorless, self-righteous True Believers with their one true path never understood what Ed was saying.

Two decades ago, I lived for a few years in Zuni Pueblo. Out in the slickrock breaks and piñon-juniper forest of western New Mexico, the sunsets break down the walls René Descartes built us. Here his separations between mind and body, Man and Nature, dissolve.

At Zuni ceremonials I learned more than I had in all the Sundays I spent in church as a kid. Most of all, I learned from the Mudhead Kachinas. While the most sacred rituals were being performed, the Mudheads were cutting up, making fun of everyone—as if a nun were mooning the faithful while the Pope gave his Easter blessing. This would be horrifying to the good Moslem or Catholic or Baptist, but it is perfectly natural to the Zuni.

It wasn't until after Ed's death that I flashed back to my Zuni days, to a frozen Shalako night in early December, to the Mudheads.

Ed Abbey was the Mudhead Kachina of the conservation movement, perhaps of the whole goddamned social change movement in this country. He was Coyote. Farting in polite company. Enraging pompous prudes, prigs, and twits. Goosing the True Believers. Pissing on what was politically correct.

And thereby doing sacred work.

Ed understood deeply the need for balance. He wrote, "Be as I am . . . a part-time crusader, a half-hearted zealot. . . . It is not enough

to fight for the WEST; it is even more important to enjoy it."
Whenever we are overworked and overwhelmed, whenever we lose
our balance and our perspective, we need to read that wise advice
from Abbey.

But how does this Coyote scratching in a dusty wash apply to the
real world of environmental activism?

To begin, we need to laugh at ourselves more. We need more
humor in environmental publications; we conservationists need to
take ourselves less seriously. We need to encourage Mudheads and
Coyotes in our midst.

All camps in the environmental movement need to better accept
other camps, other approaches. Each of us is doing important work.
We each need to realize that our particular group is not the entire
environmental movement.

All of us toiling in the fields of environmental, peace, and justice
causes need to take our philosophies, our worldviews, with a grain
of salt. The path I follow—Deep Ecology, the idea that wilderness
has value in and of itself, that a Grizzly Bear has just as much right
to her life as a human has to hers—is not a perfect, cast-in-concrete
dogma. If ever it becomes that, it will be worthless—just another
rigid gospel. Nonetheless, we absolutely need a mythology to guide
us in our work. (We are, after all, only human. Let us not deny
anthropology.) My mythology and that of my associates is Deep
Ecology, or biocentrism. But no matter how valid it is, how deep
it is, we must constantly acknowledge that it is still an abstraction.
It is a good, workable basis by which to operate. But it is not
infallible scripture. Like any abstraction, like any articulation of
reality, the Deep Ecology philosophy is just a map. And the map is
never the territory.

Finally, we *do* need to take Cactus Ed's advice: Run those rivers,
climb those mountains, encounter the Griz. . . .

And piss on the developers' graves.

17 | Dreaming Big Wilderness

Leave it as it is. You cannot improve on it. The ages have been at work and man can only mar it.

—Theodore Roosevelt

In 1963, Martin Luther King, Jr., stood before the nation in Washington, D.C., and cried out, "I have a dream!" Riveted by his eloquence, we took a significant step toward realizing his dream of racial justice one year later, when President Lyndon Johnson signed the Civil Rights Act.

The summer that the Civil Rights Act was signed into law, another controversial measure, also requiring a courageous campaign of many years, became law. This legislation embodied a dream no less noble or eloquently stated than Dr. King's:

In order to assure that an increasing population, accompanied by expanding settlement and growing mechanization, does not occupy and modify all areas within the United States and its possessions, leaving no lands designated for preservation and protection in their natural condition, it is hereby declared to be the policy of the Congress to secure for the American people of present and future generations the benefits of an enduring resource of wilderness.

The year 1989 marked the twenty-fifth anniversary of the Wilderness Act. As was appropriate, celebrations were held around the United States by the Forest Service (which had stoutly opposed the legislation three decades ago) and by conservation groups ranging from the national Sierra Club to Friends of the Gila Wilderness in New Mexico. The Wilderness Society published commemorative articles in their magazine *Wilderness,* and produced a special map of the National Wilderness Preservation System. Now that the much-deserved celebration is over, it would be good to consider the implications of the Wilderness Act more deeply. It is time to ask whether the results have lived up to the dream. And, more important, we must ask what dreams we have for the twenty-five years ahead.

The farthest and clearest views come after a long, sweat-soaked slog uphill. As we sit catching our breath in the sunny air of this flower-spangled high divide, sipping cool mountain water from a spring, let us look back down the winding trail and trace our pathway to this point. And then, gazing in the other direction, let us study the steep slopes and precipitous canyons ahead and plan a route to our further destination.

The dream of congressional protection of Wilderness was the product of many men and women, but two who would not live to see it realized—Bob Marshall and Howard Zahniser—were perhaps most responsible for it.

Bob Marshall called Aldo Leopold "the Commanding General of the Wilderness Battle." Without discounting Leopold's importance, Marshall's battlefield description befits himself better. Leopold was responsible for the first Wilderness Area administratively established by the United States Forest Service (New Mexico's Gila in 1924); but it was Bob Marshall (who originally organized and funded The Wilderness Society in 1935), as Head of Recreation for the Forest Service in the late 1930s, who instituted a *system* for protecting wildlands on the National Forests. Regulations promulgated by Marshall designated about 14 million acres as "Primitive Areas" and set up a process to study them more formally and establish firm boundaries. After such study, each would be designated a "Wilderness Area" if over 100,000 acres, or a "Wild Area" if under

100,000 acres. Two months after his proposed regulations were approved by Forest Service Chief Ferdinand A. Silcox, Marshall, a legendary long-distance hiker, died on a train of a heart attack at the age of 38 in November 1939.[1]

After World War II, the Forest Service continued the studies required by Marshall's regulations, but by this time the agency was in thrall to timber production. A new generation of conservationists viewed with alarm the dismembering of Marshall's system as the Forest Service proposed smaller Wilderness Areas than the original Primitive Areas. In 1951, Howard Zahniser, executive secretary of The Wilderness Society, called for congressional designation of Wilderness Areas. Zahniser drafted a bill and began to organize around his dream. In 1955, Senator Hubert Humphrey and Representative John Saylor introduced the first version of the Wilderness Act. After eight years of hearings in Washington and the West, after dozens of revisions to answer industry and Forest Service objections, and after dogged shepherding by "Zahnie," the Wilderness Act was signed into law on September 3, 1964, by President Lyndon Johnson. Accepting one of the pens used to sign the bill was Zahniser's widow, Alice. Zahnie had died of a heart attack three months before.

The Wilderness Act established a National Wilderness Preservation System initially composed of the preexisting Wilderness and Wild Areas on the National Forests. It directed the Forest Service to study and make recommendations on the remaining Primitive Areas, and directed the National Park Service and U.S. Fish and Wildlife Service to do likewise for roadless areas on the lands they managed. Congress reserved for itself the power to add or delete areas to or from the system. Construction of roads, use of motorized vehicles or equipment, and commercial cutting of timber were prohibited in Wilderness Areas.

1. A fine biography of Marshall is James M. Glover's *A Wilderness Original: The Life of Bob Marshall* (Seattle: The Mountaineers, 1986). Excellent histories of the wilderness preservation movement are Roderick Nash's *Wilderness and the American Mind* (New Haven: Yale University Press, 1982 [third edition]) and *Battle for the Wilderness* by Michael Frome (New York: Praeger, 1974).

Despite Zahniser's death, wilderness-minded citizens did not rest on their laurels after passage of the Wilderness Act. They worked to rectify five omissions in the heavily compromised legislation that became apparent over the next decade: free-flowing rivers; additional National Forest areas in the West; National Forest areas in the East; wildlands managed by the Bureau of Land Management (BLM); and Alaska lands. For each of these campaigns—for Wild Rivers, additional Western National Forest lands, Eastern National Forest lands, BLM lands, and Alaska lands—worthy grassroots heirs to the legacy of Bob Marshall and Howard Zahniser pursued a dream. Nine million acres of Wilderness were established with the Wilderness Act in 1964; thanks to these heirs, 90 million are protected today.

I have no desire to be the skunk at the garden party. I am glad for every acre placed off limits to management by chain saw, drill bit, and bulldozer. The oaken hills of Ishi; the dripping giant forest of Olympic; the hungry hardwoods of Cranberry gobbling up old logging trails; the knife-edge divide of the Weminuche slicing fat, black bellies of thunderclouds; Sandhill Cranes flying dimly through an August snowstorm in Denali; the transition of Saguaro to Corkbark Fir going up Pusch Ridge; the Baldcypress swamps of Okefenokee; Elk bugling in the Gros Ventre; the surprise of a Cottonmouth in the Big Slough; the churning whitewater in the River of No Return a mile below the highlands; the wheeling of ravens in the surreal Bisti badlands—I know all of these, and I am thankful to the dedicated gang of wilderness fanatics who labored in the trenches for each of these Wilderness Areas.

Nonetheless, let us not fool ourselves. There is more land under asphalt and concrete in the lower forty-eight states than there is under Wilderness designation. Of the 90 million acres in the National Wilderness Preservation System, only 34 million are outside of Alaska. For every acre in the lower forty-eight we have protected since 1964, at least an acre of equally natural land has been lost to clearcuts and roads. In the National Forests, for every acre designated as Wilderness, almost three acres have been "released" to potential development.[2] In our heart of wildness, Yellowstone National Park, a Grizzly population that was healthy in 1964 now teeters above the precipice of extinction. The offspring of California Condors who flew free over the Sespe twenty-five years ago today blink against automobile exhaust behind bars. On the eve of the Wilderness Act, the National Forests of the West supported extensive old-growth forests. In 1990 those forests are shattered, bleeding ruins, with the scattered remnants nearly unable to function as intact ecosystems.

In the quarter-century since the passage of the Wilderness Act, conservationists have waged a struggle to preserve a *portion* of the remaining wildlands in the United States. Wilderness preservation groups have not asked for protection of *all* roadless or undeveloped areas, even though these lands amount to only 8 percent of the total land area of the United States outside of Alaska. Even for the areas discussed above—Wild and Scenic Rivers, additional Western National Forest roadless areas, Eastern National Forest roadless areas, BLM roadless areas, and new Parks and Refuges in Alaska—compromise quickly weakened the original vision. In the Forest Service's second Roadless Area Review and Evaluation (RARE II) in 1978, the Sierra Club, The Wilderness Society, and their allies asked only that 35 million of 80 million roadless acres on the National Forests be protected. It has been a similar story with the holdings of the Bureau of Land Management studied for possible Wilderness recommendation; and although the Alaska National Interest Lands Conservation Act stands as the outstanding conservation achievement of the 1970s, even in Alaska environmental groups never considered proposing that development be frozen and that all of the state's wildlands remain undeveloped.

It has been taken for granted that the implacable forces of indus-

2. National Forest Wilderness bills in the 1980–1984 period designated 8.7 million acres of Wilderness and released 19.9 million acres to development. But of the 8.7 million acres designated, over 1 million acres had been previously protected as Primitive Areas, so in reality only 7.5 million acres of formerly unprotected National Forest land was being protected.

trialization will continue to conquer the wilderness. Environmentalists, as reasonable advocates within the mainstream of modern society, have gone out of their way to appear moderate and willing to
compromise. We have acquiesced in the clearcutting of ancient
forests, in massive road building schemes on our public lands, in
mineral and energy extraction in pristine areas, and in the destruction of "problem bears." We have accepted that some wildlands will
be—and should be—developed. We have merely asked that some of
these—generally the scenic ones—be spared.

With few exceptions, dreams have been replaced with political
pragmatism. "We live in a world of decisions," we are told, "and
we must be ready to deal with the people who make decisions."
Vision has fallen by the wayside as we have become mired in the
"political quag" John Muir dreaded. Meek reaction to bureaucratic
initiatives has come to the fore.

In short, wilderness conservationists have lacked a comprehensive
vision since the passage of the Wilderness Act. In most cases we
have simply responded to agency programs. We've fought brush-
fires but have failed to articulate and campaign for a representative
National Wilderness Preservation System worthy of the name. We
have accepted the dominant social paradigm, the inevitability of
continued industrialization and development of open spaces. We
have had no dream for such noble but vanishing species as the
California Condor, the Grizzly Bear, the Gray Wolf. We try to
hang on to their diminishing habitats, their puny populations as
museum pieces, but not as growing, vigorous, living parts of the
functioning world.

If the Wilderness System is to be anything more than a museum
offering a tantalizing glimpse of a bygone North America, if it is to
be more than an outdoor gymnasium and art gallery, if it is to
preserve representative samples of dynamically evolving natural
ecosystems, we must have an inspirational objective instead of ob-
sequiously accepting what crumbs are tossed to us by Louisiana-
Pacific, the Forest Service, Senator Mark Hatfield, and Exxon. Conservationists must lead, instead of politely responding. We must ask
deeper questions of our nation: Is 2 percent of the acreage of the

forty-eight states adequate for our National Wilderness Preservation System? Are twenty-five condors sufficient? Six hundred Grizzly Bears? A few minuscule remnants of the old-growth cathedral forests of Oregon?

Have we logged too much virgin forest? Have we built too many roads? Have we dammed too many rivers? Have we driven the Griz, the Gray Wolf, the Cougar, the Bighorn, the Bison from too many places? Have we drained too many wetlands? Were the exterminations of the Passenger Pigeon, Sea Mink, and Heath Hen, and the plowing of the Great Plains all monstrous mistakes?

Are not Wilderness Areas the world of life: vibrant ecosystems where natural processes still reign, and evolution runs its course?

If we fail to ask these deeper questions of the nation, if we neglect to proclaim a magnificent and noble dream as did Martin Luther King, Jr., then the wilderness crusade is lost. Remnants of the wild, with truncated flora and fauna, will haunt future generations as the shadows of what once was real.

From the beginning, preservationists have ducked the hard questions. We have deftly avoided admitting our real reasons for wilderness preservation. In championing National Parks and Wilderness Areas, we have allowed ourselves to become trammeled in the values of Babbitt. The original advocates of National Parks—those promoting Yosemite and Yellowstone—wanted to preserve not natural diversity, not wilderness, but simply spectacle, the curiosities of nature. Alfred Runte, in his book *National Parks: The American Experience,* [3] terms their reasoning "monumentalism." Proponents of the Parks also used the argument of "worthless lands." Our nation could afford to set aside Yellowstone Park, they said, because it was unsuited for agriculture. Moreover, any minerals were buried out of reach beneath lava. Even preservationists like John Muir frequently fell back on monumentalism and worthless lands, just as lobbyists for the Sierra Club do today to justify new Parks and Wilderness Areas or to excise "controversial" portions (i.e., those with trees) from proposals.

3. Lincoln: University of Nebraska Press, 1987 (second edition, revised).

The other popular arguments initially used for preservation were based on utilitarian and primitive recreational values. One hundred years ago, the state of New York set aside state lands in the Adirondacks as "forever wild" not for the inherent goodness of wildness but to protect the watershed for booming New York City. The first Primitive Areas on the National Forests were established in reaction to Henry Ford's pervasive machine—foresters who had grown up with pack saddles and diamond hitches were loath to see those pioneer skills lost. Since then, champions of wilderness have emphasized the values offered human beings, watershed protection and primitive outdoor recreation being the two most commonly cited.

This is why we end up with biologically impoverished lands above timberline in our preserves, why the new Parks in Alaska ("Where we have our last great chance to do it right from the beginning" as the Alaska Coalition proclaimed) came out as gerrymandered by commercial interests as were their siblings in the lower forty-eight. When you say you only want worthless lands, you get worthless lands. Monumental, yes. Scenic, indeed. Even breathtaking. But not the rich, virile areas needed by sensitive species. Yellowstone National Park, as spellbinding as it is, cannot stand on its own as an ecosystem. Its boundaries were drawn not by nature but by politicians who kowtowed to the dollar. There is a reason why even the smoking guns of Montana meat hunters cannot persuade the Bison herd to stay in the Park. Bison can't eat scenery or geysers. The unspectacular private lands north of Yellowstone, now used as cattle ranches, are their necessary winter range—as they have been for millennia.

Real wilderness is far different from that which forms our current National Wilderness Preservation System. Most areas in the system are small enough to cross on foot in a day,[4] and almost all have lost

---

4. A rough calculation reveals that only 4 percent of the units of the National Wilderness Preservation System are 500,000 acres or larger, while 34 percent are under 10,000 acres in size. An area of 500,000 acres is equivalent to a square twenty-seven miles on a side; 10,000 acres is equivalent to a square less than four miles on a side. The center of most Wilderness Areas is not more than three miles

important members of their original fauna. To Aldo Leopold, a wilderness was an area large enough for a two-week pack trip without crossing your own tracks. To Grizzly Bear cinematographer Doug Peacock, an area is wilderness if it contains something bigger and meaner than you—something that can kill you. Lois Crisler wrote in *Arctic Wild*, [5] "Wilderness without animals is dead—dead scenery. Animals without wilderness are a closed book."

Thoughtful biologists and conservationists have come to understand in the last twenty-five years that the destruction of Earth's natural diversity is caused not by the mere excesses of industrial civilization but by the inherent attributes of that society: overconsumption, overpopulation, and our notion of mastery over nature. They now realize that designated Wilderness Areas and National Parks cannot survive as effective sanctuaries if they remain island ecosystems, that habitat islands in a sea of development will lose key species (those that require larger territories to maintain sustainable breeding populations). They have sadly acknowledged that outside impacts such as acid precipitation, other forms of air pollution, and toxic and radioactive contamination can devastate the natural integrity of protected areas, that no preserve is immune from the fouling of Earth's air, water, and soil by industrialism. And, with horror, they are beginning to recognize that global impacts such as the greenhouse effect and depletion of the atmospheric ozone layer will play havoc with all ecosystems worldwide, including those in sanctuaries. Minor reform of our economic system and better stewardship will not safeguard the incredible diversity of life hatched by nearly four billion years of evolution. The long-term protection of natural diversity and the processes that sustain it will require fundamental changes in the role we humans play on our planet.

A vital part of grappling with these formidable problems is to envision and promote a National Wilderness Preservation System

as the raven flies from the boundary, and the farthest point from a road in the lower forty-eight states is a mere twenty-one miles.

5. Lois Crisler, *Arctic Wild* (New York: Ballantine Books, 1973).

in the United States that is truly national and representative, and that preserves native diversity. By clearly stating a dream of ecological wilderness and campaigning for it in the national arena, we would come much closer to safeguarding real wilderness than we would if we continued to fight only for the traditional backpacking parks, open-air zoos, and scenic preserves. Another benefit of such a program is that the very process of proposing and working for ecological wilderness may be the most effective means of redefining the role of humankind in nature; it may be the best way to bring about the change of consciousness that will, in Aldo Leopold's words, transform "the role of *Homo sapiens* from conqueror of the land-community to plain member and citizen of it." Such a reformation of our role would enable us to transform our gluttonous lifestyle, which causes acid rain, the greenhouse effect, and depletion of the ozone layer. And if the materialistic society of the United States can find the humility to establish substantial nature preserves, we will at last set an example for other nations, particularly those in tropical regions where native diversity is especially abundant *and* imperiled. How can we lecture Brazil to cease the destruction of the Amazonian rain forest while we shred the library of ecological richness found in the ancient forests of the Pacific Northwest? How dare we enjoin starving tribespeople of East Africa from slaughtering the great herds, when we cannot find the generosity to give the Bison, Gray Wolf, and Grizzly the range they need?

Constructing a meaningful but politically possible National Wilderness Preservation System requires us to outline our goal carefully, and the steps to achieve it. I offer the following blueprint.

Draw the line on what is now wild. Not one more acre of old-growth or substantially natural forest should be cut. Not one more mile of new road bladed into a roadless area. Not one more Grizzly Bear murdered in Yellowstone. Not one more free-flowing river dammed. Too much has already been lost.

Recover native ecosystems. In many cases, to recover native ecosystems, to reintroduce extirpated wildlife, and to repair damaged landscapes, all that is necessary is to close roads, cease damaging activities, and leave the land alone. In others, minor hands-on restoration may be required: physically reintroducing extirpated species, removing a few developments, and performing minor watershed rehabilitation. Some areas will require more expensive, long-term, and active management to be returned to a state of natural wildness. These areas should be designated Wilderness Recovery Areas, with more intensive rehabilitation work allowed, until wildness is restored.

Restore large ecological wilderness preserves east of the Rockies.

■ A 10- to 20-million-acre Great Plains National Park with free-roaming Bison, Elk, Pronghorn, Grizzly, and Gray Wolf.

■ A 5-million-acre North Woods International Preserve around the Boundary Waters Wilderness in Minnesota and Quetico Provincial Park in Ontario.

■ A large deciduous forest Wilderness Recovery Area in the Ohio Valley with Elk, Bison, Gray Wolf, and Eastern Panther.

■ A 10-million-acre National Park in northern Maine with Gray Wolf, Lynx, Wolverine, and Woodland Caribou.

■ A 1.5-million-acre Bob Marshall Greater Wilderness in the Adirondacks of New York, with Gray Wolf and Eastern Panther.

■ A 4-million-acre Wilderness Recovery Area in the Southern Appalachians, centered around Great Smoky Mountains National Park, with Eastern Panther and Elk.

■ A 5-million-acre Everglades/Big Cypress National Park in Florida.

These core areas and smaller Wilderness Areas and Wilderness Recovery Areas should be linked to one another by undeveloped corridors. Such corridors are vital for the transmission of genetic diversity between core preserves. Without such corridors, preserves become ecological islands, and populations of low-density species, such as large predators, may become inbred. When it is determined that suitable habitat exists, extirpated species should be reintroduced if it appears unlikely that they will return to the area on their own. The near extinction of mature American Chestnut trees (due to an exotic disease) leaves a gaping hole in the Eastern forest. The Forest Service and National Park Service should fund a research project to develop a blight-resistant American Chestnut that could be reintroduced to its former habitat in these protected areas.

Restore major roadless areas in the West. There are currently thirty-eight areas where minor road closures would create core roadless areas of more than a million acres:

Area	State(s)	Acreage (millions)
North Cascades	Washington	3
Olympic Mountains	Washington	1.2
Kalmiopsis/Siskiyous	Oregon/California	2
Hells Canyon/Eagle Cap	Oregon/Idaho	1.5
Selway/River of No Return	Idaho/Montana	5.5
Great Rift	Idaho	1
Owyhee	Idaho/Oregon/Nevada	8
Oregon Desert	Oregon/Nevada	3
Bob Marshall	Montana	3
Beartooth	Montana/Wyoming	1.5
North Absaroka	Wyoming	1
Upper Yellowstone/ South Absaroka	Wyoming	2.5
Tetons/SW Yellowstone	Wyoming/Idaho	1

Area	State(s)	Acreage (millions)
Wind Rivers	Wyoming	1.2
Red Desert	Wyoming	1
Maroon Bells	Colorado	1
San Juan Mountains	Colorado	1.5
Desolation Canyon	Utah	2.2
High Uintas	Utah	1
Canyonlands	Utah	3
San Rafael/Wayne Wonderland	Utah	1
Escalante/Kaiparowits/ Henry Mts.	Utah	3
Desert Game Range	Nevada	1.5
Black Rock Desert	Nevada	2.5
Smoke Creek Desert	Nevada/California	1
High Sierra	California	3
Yosemite North	California	1
Los Padres	California	1.5
Inyo/Saline/ Cottonwood	California	2
Panamint Mountains	California	1.5
Mojave Desert	California	1
Bill Williams River	Arizona	1
Kofa	Arizona	1.5
Cabeza Prieta	Arizona	2
Galiuro/Pinaleno	Arizona	1
Grand Canyon/Kaibab	Arizona	3
Gila/Black Range	New Mexico	1.5
Guadalupe Escarpment	New Mexico/Texas	1

These million-acre or larger wilderness units should then become the cores for even larger wilderness complexes linked to each other and to smaller Wilderness Areas by wild corridors.

Reestablish native species. The Grizzly will not survive restricted to the dwindling Yellowstone and Bob Marshall/Glacier ecosystems. Populations should be reestablished in the Gila Wilderness of New Mexico, the Blue Range of Arizona, the Weminuche/South San Juans of Colorado, the High Uintas of Utah, the Kalmiopsis of Oregon, the Marble Mountains and Siskiyous of California, the North Cascades of Washington, and Central Idaho. The Gray Wolf should be returned to these areas and others. A million and a half acres in the Los Padres National Forest, northwest of Los Angeles, should be totally closed to human use or entry in order to protect the California Condor after reintroduction. In suitable areas of southern New Mexico, Arizona, and Texas, the Jaguar, Ocelot, and Jaguarundi should be reintroduced. Bighorn Sheep, Bison, Pronghorn, River Otter, Woodland Caribou, and other once-widespread species should be widely propagated in their former habitats.

Terminate commercial livestock grazing on the Western public lands. Only 3 percent of our nation's red-meat supply comes from public land, and the government spends more on managing this private grazing than it receives in fees from the grazing permittees. Grazing has been the single most important factor in the devastation of intermountain ecosystems: the widespread decimation of bear, wolf, Mountain Lion, Elk, Pronghorn, Bighorn, and Bison; destruction of native vegetation; and severe damage to watersheds and riparian systems.

Rehabilitate free-flowing rivers. Perhaps more than any other ecosystem type in the United States, rivers and riparian habitats have been abused, altered, and destroyed. High priority should be given to rehabilitating free-flowing rivers, eliminating disruptive exotic fish species, and restoring native fish and other riverine species where feasible. Not only should no new dams be built, but a program should be launched to remove dams and recover free-flowing rivers.

Discard the notion of static landscape preservation. What is being preserved in Wilderness Areas is the process of evolution, of speciation, of seral changes in ecosystems. Natural landscapes should be large and diverse in order to absorb catastrophic events such as huge forest fires, insect and disease outbreaks, temporary regional extinctions, and cyclical population fluctuations. (In a large enough preserve or complex, a certain habitat may be wiped out by a stochastic[6] event such as the 1988 Yellowstone fires, but similar habitats will continue to exist elsewhere in other parts of the area, or in other areas connected by corridors.) Wilderness proponents need to learn from conservation biologists, who in turn need to see grassroots conservation activists as their natural allies and the management of public lands as a vital opportunity.

Preserve wilderness for its own sake. Conservationists must develop a new (old) reason for wilderness, a new understanding of the place of humans in the natural world, a new appreciation for the other nations inhabiting this beautiful blue-green living planet. We should recognize that the true reason we favor wilderness preservation is *Wilderness for its own sake.* Because it's right. Because it's the real world, the arena of evolution; because it's our home. The Gray Wolf has a claim to live for her own sake, not for any real or imagined value she may have for human beings. The Spotted Owl, the Wolverine, Brewer's Spruce, the fungal web in the forest floor have the following of their own intertwined evolutionary paths as their due. Not only should conservationists recognize that it is the inherent value of natural diversity that argues for its preservation in our hearts, *but it is also the most effective argument for preservation;* we should state that rationale forthrightly to the public. Unless we

6. *Stochastic* is used by ecologists to indicate a random alteration in an ecosystem, brought about by fire, hurricane, or other natural perturbation.

challenge our fellow humans to practice self-restraint, to share Earth voluntarily with our wild fellows, the wilderness crusade is pissing in the wind.

Why does a man with a lifespan of seventy years think it proper to destroy a two-thousand-year-old redwood to make picnic tables? To kill one of thirty breeding female Grizzlies in the Yellowstone region because she ate one of his sheep? To rip through a five-thousand-year-old Creosote Bush on a motorized tricycle for some kind of macho thrill? To dam Glen Canyon and Hetch Hetchy for electricity and water to irrigate lawns?

Until we learn to respect these others as our equals, we will be strangers and barbarians on Earth. Wilderness—*real* wilderness—is the path home. The articulation of that truth is the vital duty of the preservation movement. We cannot achieve it by hiding behind the anthropocentric arguments of monumentalism, worthless lands, utilitarianism, or primitive recreation. We can do it only by stating what we truly believe, and challenging humankind with that ethical ideal.

18 | Making the Most of Professionalism

There is just one hope of repulsing the tyrannical ambition of civilization to conquer every niche on the whole earth. That hope is the organization of spirited people who will fight for the freedom of the wilderness.

—*Robert Marshall*

Far too many hours of my early life were wasted listening to country-western music. I sat in countless bars, sipping beers, listening to the jukebox, and enjoying a sweet sort of melancholy. Across endless empty spaces of the American West, I played with the radio dial in my truck, trying to coax in that fadin'-in, fadin'-out, country station in Silver City or Boise or Casper, all the while savoring a languorous loneliness and a lukewarm six-pack.

The bittersweet tears falling in those bygone beers were for old loves of the feminine kind, but if I were shedding them today, they might be for a love of another sort.

Country music, as every listener knows, has as its essential theme a love broken by the lure of the bright lights. Innumerable songs have been recorded by Hank and Merle and Dolly and Emmy Lou and Charlie and Tammy and George and all the rest about lovers from the country who are torn apart when one of them is drawn to the city. The purity, innocence, sweetness, and strength of nature is lost, as the wayward lover—the barefoot boy, the freckle-faced girl—is lured by the fast life of the city. It's the oldest story in civilization, at least as old as that peasant boy or girl who first saw the glow of Sumer eight thousand years ago.

The love for whom I mourn is a cause, an ideal, an ethic, a movement—the tribe of Bob Marshall's spirited men and women.

Yeah, that old love of mine—and I knew her back when—is in the fast lane now. She's got style that puts a country bumpkin like me to shame. She's in high society, for sure.

But can she still hear the birds sing? Does she ever look at a sunset? Does she still like to feel the mud squish between her toes?

Hmmm. Better give me another Lone Star. Yeah . . . and here's a quarter for the jukebox. . . .

I don't listen to country music anymore. Even the music itself has gone to the city and gotten duded up. I don't cry over lost girl-friends either. I'm old enough to be married—and even satisfied by it. And I don't drink much in bars now. Being a respectable, middle-aged, middle-class bald guy, I get plastered in my backyard.

During the early 1980s, as Johnny Sagebrush and I toured the country with the original Earth First! Road Show, criticism of the conservation movement was a key part of my message. Not even my advocacy of monkeywrenching has caused such hard feelings toward me within the Sierra Club, The Wilderness Society, and other mainstream groups as that criticism has.

I've been haunted by old friends who still work for such groups. I wake up in the middle of the night to see them standing by my bed with hurt in their eyes, pointing accusing fingers at me. I feel guilty then, and disloyal. Perhaps some of my criticism of the "Gang of Ten" was harsh, insensitive, even excessive.[1] I still belong to

1. The Gang of Ten is a term for the ten largest mainstream environmental organizations—the conservation establishment, if you will. These ten groups are The Wilderness Society, Sierra Club, National Audubon Society, National Wildlife Federation, Natural Resources Defense Council, Environmental Defense Fund, Defenders of Wildlife, National Parks and Conservation Association, Izaak Walton League, and Environmental Policy Institute. (EPI has recently merged with Friends of the Earth and the Oceanic Society.)

several of these groups and I support most of what they do; nevertheless, I feel that noteworthy problems exist and are growing even as the groups grow. Many of these problems can be lumped under the umbrella of "professionalism." So let me treat you to two or three six-packs of entirely practical solutions. I offer my criticisms and suggestions as an insider—as someone who has worked full-time on wilderness for twenty years—and I offer them constructively and with respect. Marshall's "organization of spirited people who will fight for the freedom of the wilderness" is my home, my family, my clan. I want it to remain spirited, and I want it to fight for the freedom of the wilderness as effectively as possible.

Before rasslin' these alligators, I need to lay the groundwork with a little history of the environmental movement, a bit of autobiography about my initiation into that movement, and scattered shards of evidence that bureaucratic problems are undermining the fighting spirit for which Marshall called.

Contrary to common belief, the effort to protect the "environment" (a word I dislike, but we seem to be stuck with it) did not begin with Earth Day 1970. The first Earth Day is, however, a handy marker for the transformation of the conservation movement into the environmental movement—with consequences both good and bad.

Prior to 1970, the conservation movement concerned itself primarily with the establishment and management of National Forests, National Parks, and Wilderness Areas; protection of wildlife habitat and establishment of wildlife refuges; development of game-protection laws and standards of good sportsmanship for hunters and fishers; prevention of soil erosion; and fighting dams on free-flowing rivers. Members of conservation groups were generally outdoorspersons—hunters, fishers, bird-watchers, hikers, canoeists, mountain climbers, and naturalists.

The interaction of two factors by the late 1960s transformed conservation into environmentalism. First, several books brought environmental problems to the public's attention. Rachel Carson's *Silent Spring* (1962) dramatized concerns about the use of pesticides and herbicides at a time when worry about air and water pollution was growing. Secretary of the Interior Stewart Udall's *The Quiet Crisis* (1963) was a widely popular, albeit superficial, book about the conservation movement. Paul Ehrlich's *The Population Bomb* (1968) portrayed the threat of overpopulation in lurid, prophetic terms. Barry Commoner's *The Closing Circle* (1971) followed *Silent Spring* [2] with warnings about our increasingly toxic environment.

Second, the demographics of those concerned with environmental issues changed. *Silent Spring, The Quiet Crisis, The Population Bomb,* and *The Closing Circle* reached millions of average citizens who were not necessarily outdoorspersons—as did David Brower's ambitious and farsighted publishing program of large-format Sierra Club photo books. Also, by 1970, many young people with experience in the antiwar and civil-rights movements discovered "ecology" and became active on a variety of issues.

The concern of conservationists with wildlife and wildlands had generally not entailed questioning the overall problems of industrial society. To a certain degree, conservation had operated in a vacuum, apart from other social and political questions. *Silent Spring* and the others of its genre created a more holistic picture, tying wildlife and other "proper" conservation issues to the general problems of society. Ironically, even though this was a far more ecological approach, and although Carson, Ehrlich, and Commoner were biologists (and Carson is hailed by Bill Devall and George Sessions in *Deep Ecology* [3] as an early biocentric thinker), the general influence of this trend was to make conservation—now environmentalism—more anthropocentric.

Moreover, the young people who entered the movement around Earth Day were not necessarily interested in the back of beyond or nature, except in an abstract sense. They were oftentimes more concerned with the impacts of air and water pollution on human

2. Rachel L. Carson, *Silent Spring* (Boston: Houghton Mifflin, 1962). Stewart L. Udall, *The Quiet Crisis* (New York: Holt, 1963). Paul R. Ehrlich, *The Population Bomb* (New York: Ballantine, 1968). Barry Commoner, *The Closing Circle: Nature, Man, Technology* (New York: Knopf, 1971).

3. Bill Devall and George Sessions, *Deep Ecology: Living As If Nature Mattered* (Salt Lake City: Peregrine Smith Books, 1985).

beings, and with the role of large corporations in causing environmental problems just as corporations caused social problems and contributed to the military-industrial complex. These newcomers to conservation called for "relevancy" and tried to draw the focus away from wildlife and wildlands and toward urban environmental issues.[4]

The young converts to environmentalism also brought a confrontational style and an anticapitalist bent with them. Previously, conservation groups had avoided confrontation whenever possible, preferring instead to work things out with government officials like the Chief of the Forest Service or the Director of the National Park Service. One of the key reasons for the 1969 firing of David Brower as executive director of the Sierra Club was a negative reaction on the part of the Club's old guard to his increasingly confrontational approach with government agencies and polluting corporations. Conservation was not traditionally couched in liberal or leftist terms. Until the mid-1970s, environmental issues were clearly bipartisan, with some of the strongest supporters for the movement coming from the Republican Party.

It may well be debated whether allying environmentalism with liberalism in general was a positive change. But that is a topic deserving analysis on its own. I'm teasing a different rattlesnake here.

Although environmentalism was launched by a more holistic approach and a better understanding of the science of ecology, it unfortunately resulted in making the former conservation movement less nature-centered and more human-centered. Elsewhere I discuss the need to make wilderness the keystone of conservation

4. This discussion is highly generalized. Many people drawn to the environmental movement in the late 1960s and early 1970s *were* outdoor-oriented, and the boom in backpacking and river-running played a significant role in attracting newcomers to conservation organizations. What I am pointing out are trends and tendencies that may explain some of the changes in the movement over the last several decades. This chapter is written in generalities, and there are numerous specific exceptions to everything I say. I am merely showing trends to illustrate my point.

once again, and to take a more biocentric approach to ecological issues. Here, however, I would like to turn my flashlight more specifically on the transformation of conservation organizations themselves. Before I go into that, though, I'd like to place my views in context by describing how I became involved with the wilderness preservation cause.

My favorite books, even before I could read, were two volumes from the American Museum of Natural History: *American Wildlife Illustrated* and *Wildlife the World Over Illustrated.* I remember my mother reading them to me when I was five years old; I still have them in my library. Unfortunately, my family was not one of bushwhackers; a picnic in the Sandia Mountains was about the closest we came to a wilderness experience. I could scarcely wait to turn eleven and join the Boy Scouts. I completed the requirements for Eagle Scout at the age of thirteen, and went on my first backpack trip, in Washington's Glacier Peak Wilderness Area, in 1962 with my Scout troop. That trip made me a wilderness fanatic.

After Earth Day 1970, Santa Fe became a focus for radical environmentalism with the formation of an eco-anarchist group called Black Mesa Defense, organized by Jack Loeffler and others, that supported Navajos fighting a massive coal strip-mining project on the reservation. Through mutual friends, I met some of the Black Mesa Defense people and began to drop in as a volunteer—to lick stamps and stuff fliers in envelopes. Since I was a whitewater boater, I was particularly interested in an offshoot of Black Mesa Defense called River Defense.

As a backpacker, I had become vaguely aware of the Wilderness Act. Activated by the Santa Fe crowd, I wrote to the Forest Service, asking about potential new Wilderness Areas in New Mexico. The Forest Service responded that they were launching a program to look systematically at potential new Wilderness Areas—the Roadless Area Review and Evaluation (RARE). They suggested I contact the New Mexico Wilderness Study Committee to find others who were working on the issue. I wrote to the Study Committee, received a rather tentative and aloof reply, and started looking for blank spots on topographic and Forest Service maps.

The New Mexico Wilderness Study Committee finally invited me to a wilderness workshop in March of 1972. The workshop featured Harry Crandell and Jerry Mallett from the Washington, D.C., and Denver offices of The Wilderness Society, respectively. The primary focus of the workshop was the Forest Service's RARE program. The National Forests of New Mexico were parceled out to various grassroots "honchos" to organize committees to investigate potential new Wilderness Areas and develop support for protection. Since no one else volunteered, I offered to be honcho for the Gila National Forest in southwestern New Mexico, and was reluctantly accepted by the old guard of the Study Committee.

While working on the Gila National Forest's roadless areas, I learned that the Forest Service was studying the Gila Primitive Area for possible addition to the Gila Wilderness Area. Hearings were planned for late fall of 1972. Under the tutelage of an experienced Silver City conservationist, Jim Stowe, I took on the project of coordinating the combined conservationists' response to the Forest Service study, and of organizing turnout at the hearings.

My work on the Gila Primitive Area reclassification study drew the attention of The Wilderness Society, and I was invited to a week-long Washington, D.C., lobbying seminar in January 1973. After the seminar, Harry Crandell asked me if I would be interested in a job as their field consultant for New Mexico. I was sent home through Denver, Colorado, where Clif Merritt maintained the Western Regional Office for TWS. Clif gave me a postdoctoral short course on the finer points of the Wilderness Act and the fundamentals of conservation organizing. I was to be paid a salary of $250 a month, plus fifty dollars a month for expenses. Not much, but it beat shoeing horses, and I would now be paid to work full-time on wilderness preservation.

I spin this tale of my hiring as a field consultant for two main reasons: It is representative of how many activists came to work for environmental groups in the early 1970s; and knowing our own history, as well as the history of the transformation of the conservation movement into the environmental movement, is important for understanding some of the current problems of the movement and for devising possible solutions.

One of the distinguishing characteristics of the environmental movement since the first Earth Day in 1970 has been the increasing number of full-time, paid staff members of conservation groups. Professional staff, like myself, have contributed much to the successes of the cause and are a vital force in the battles ahead. Nevertheless, a number of problems can be attributed to this increasing professionalism, chief among those problems being the concentration of strategic and tactical decision-making in a small elite, and an emphasis on pragmatic politics instead of on ethical fundamentals. While the number of professional staff in the environmental movement today is unprecedented, the problems are not.

As Stephen Fox points out in his book, *John Muir and His Legacy: The American Conservation Movement*,[5] the history of the conservation movement, from the late 1890s on, can be told as the struggle between bureaucratic professionals and "radical amateurs" (enthusiastic volunteers who saw a problem and worked on it without pay).

Eighty or ninety years ago, government agencies like the Forest Service were leading elements of the conservation movement. It was Gifford Pinchot, as the first Chief of the United States Forest Service, who encouraged professionalism and, early in this century, took the conservation movement away from radical amateur John Muir. But even Pinchot would be shocked to see how his idea of professionalism has grown. In the Forest Service, the leathery-faced ranger on horseback has been replaced by the computer technician in an air-conditioned, windowless office. The symbol of the Forest Service is no longer Smokey the Bear decked out in overalls, but the Orwellian FORPLAN computer.

In citizen conservation groups, professional resource managers promoted by gun manufacturers took over after World War I: passionate outdoorsman Will Dilg was deposed as head of the largest

---

5. Madison: University of Wisconsin Press, 1985. The following discussion is based on Fox's analysis.

conservation group of that time, the Izaak Walton League; antihunting crusader William T. Hornaday was forced out of his position at the Bronx Zoo; and the National Audubon Society was neutered. One of the great chapters of conservation history, little remembered today, is the valiant struggle of volunteer activist Rosalie Edge during the 1930s to reform the Audubon Society and return it to heartfelt action. Edge anticipated the problems of professionalism with her comment, "I have met too many ironed-out, and often bitter, conservationists in professional jobs."

It is possible to combine the best of the radical amateur and the professional conservationist. From the 1930s to the 1960s, Bob Marshall, Aldo Leopold, Olaus and Mardie Murie, Howard Zahniser, Sigurd Olsen, Rachel Carson, David Brower, Stewart Brandborg, Clif Merritt, Ernie Dickerman, and Celia Hunter exemplified this union of full-time professional with hard-line enthusiastic amateur. These individuals offset the power of government agencies and corporations to dominate the movement, and they got it back on track. Since the Sierra Club's firing of David Brower in 1969, however, the environmental movement has been slowly co-opted by the concept of professionalism to the detriment of the vision, activism, ethics, and effectiveness of the cause. By the time Celia Hunter was forced out of her position as Executive Director of The Wilderness Society nine years later (1978), professionalism was well established.

I see numerous manifestations of this, as detailed on the following pages.

Many of the people who work for environmental groups today are not conservationists but technicians. As in most business, government, and citizen organizations, the technicians—accountants, lawyers, public-relations experts, and political pros—have taken over from the idealists. Where are the great conservationists leading the movement today? We look in vain for a Muir, a Leopold, a Marshall, a Murie, a Carson, or a Zahniser at the helm. What have we done with their living heirs, like David Brower or Celia Hunter? One has to look to the leaders of less mainstream groups, like the Earth First!–inspired Rainforest Action Network, the working-class Citizens Clearinghouse on Hazardous Waste, and regional or state groups like the Southeast Alaska Conservation Council and Southern Utah Wilderness Alliance to find visionaries in charge of organizations. (There are younger Browers and Hunters working for the Gang of Ten, but they are rarely in positions of management.)

Until the mid-1970s, the route to a job with an environmental group was by proving oneself first as a volunteer activist. Academic training and professional experience were worthwhile, but were not the critical considerations. The key was to be a hardworking and effective grassroots conservationist. Today the prerequisites have been reversed. It is the relevant degree or governmental experience that is important. Political operatives, not conservationists, are sought for jobs with environmental groups. It is more important, it seems, to understand the technical process of government than to feel the heart and soul of the land. (Of course, conservation professionals do need to understand the ins and outs of government, but with experience such knowledge and skills come quickly. It is more difficult to instill a passionate love for the wild in a political pro.)

Conservation groups look for potential employees who will fit smoothly into the cubbyholes of their particular organization. The businessman who took over The Wilderness Society in 1978 replaced virtually the entire experienced and effective grassroots staff of that organization with "professionals." In the Sierra Club, in at least three regions during the 1980s, the most effective, experienced, and knowledgeable conservationist in the area was passed over for a regional representative position because he did not "meet the organizational fit."

Fewer and fewer staff members of conservation groups are outdoorspersons. Many seem more comfortable on the sidewalks of Washington, D.C., or San Francisco than on the high ridges of the Cascades, Sierras, Rockies, or Appalachians; more at home in a yuppie bar than beside the campfire. This is true also of the boards of directors. Read the dusty pages of *The Living Wilderness*. The Wilderness Society Council of bygone days always met in a rustic setting next to a wilderness, and made a trip into the Big Outside afterward. Not today. Some professional conservationists today once were frequenters of the rocky trail, but have drifted away from the call of the untrammeled hills to that of Capitol Hill. Others could just as easily be working for any liberal, progressive political outfit. There are a few top staff people who are highly erudite in the lore of conservation, but they are armchair conservationists, preferring to encounter the Griz in glossy photographs on their coffee tables. Many staffers are exceptions, of course, but too few would agree with Marshall that "life's most splendid moments come in the opportunity to enjoy undefiled nature."

Staff members of conservation groups today often are career-oriented. Many see their jobs with conservation groups as stepping-stones to jobs with prominent politicians or to high positions in the next presidential administration. Too many take care not to ruffle feathers in order to preserve their opportunity to be considered later for Director of the National Park Service or Assistant Secretary of Agriculture. High salaries are commonplace. The heads of some groups pull down more than $100,000 a year (the National Wildlife Federation pays their executive director $200,000 a year). It is argued that to compete effectively in the high-priced job market of D.C. lobbying, high salaries must be offered.

Many people working for environmental groups today have a higher loyalty to the political process than to conservation. They work for conservation groups not because of an overwhelming love of wild nature but because they enjoy the glamour, excitement, and prestige of the Potomac. It is a roll-call vote in the House that mesmerizes them, not goose music. Some have strong personal loyalties to particular political figures, loyalties that often override commitment to protection of wilderness or enhancement of environmental quality. Certain politicians, like Representative Mo Udall of Arizona, are placed on a pedestal and are not intensively lobbied or criticized. (When the Tucson group of the Sierra Club refused to endorse Udall for reelection in 1988 because of his sellout on preservation of Mount Graham, the national club quickly overruled the local group.)

The viability of the group itself has become more important than the conservation mission of the group. Even Greenpeace is falling into the trap of making organizational maintenance the primary goal. There is grumbling in that once-idealistic organization as spontaneity, militancy, and flexibility are sacrificed by new strata of bureaucrats in order to gain "credibility" in the halls of power, and as an empire-building fund-raising establishment becomes the tail that wags the dog.

As organizational maintenance becomes the primary goal of a group, it begins to compete with allied groups for recognition, money, and status. Instead of trying to truly win a battle, the group merely wants to get credit for a victory, no matter how hollow it may be. Wilderness Society and Sierra Club staff receive specific directions from their supervisors to beat each other in the media race, to get more "sound bites" and "face time" than their counterparts in the other group.

A classic example of organizational maintenance transcending wilderness preservation as the goal was the public response of national groups to the designation of new National Forest Wilderness Areas in twenty-seven states from 1980 to 1984. These bills set aside 8,702,697 acres as Wilderness. Sierra Club Executive Director Michael McCloskey exulted in a fund-raising letter that "the Sierra Club had spectacular success in preserving wilderness." Other environmental spokespersons joined in the chorus. Representatives of the timber, mining, oil and gas, and livestock industries, and off-road-vehicle groups howled that far too much land was being "locked up," economic ruin would ensue, and motorized recreationists would be shut out of the National Forests. Unfortunately, none of this was true.

Left unsaid by everyone (except Earth First!) was that over a million acres of the "new" Wilderness Areas consisted of previously classified Primitive Areas, some of which had been first established in the early 1930s, thus only 7.5 million acres of the Wilderness was actually new. Most important, the Wilderness bills "released" 19,-885,600 acres of de facto wilderness to road building, clearcutting, and other destructive activities. In general, this released acreage (almost three times the size of the new Wilderness) was biologically more productive and more critical to sensitive species than the more scenic areas protected.

Although the designation of the new Wilderness Areas was important, it was not a "spectacular success" or even a victory when balanced with the released acreage. Overall, the preservation of wilderness and natural diversity suffered a defeat. Why, then, did the Gang of Ten proclaim victory, and industry defeat?

Different audiences were being targeted. The Sierra Club et al. were speaking to their members and contributors, defending their strategy and work on the bills. They feared that if they said something like, "Although we protected several important areas, wilderness preservation in general suffered a major defeat," then contributions and memberships might fall; management might even be questioned about why such a defeat was suffered. From a standpoint of organizational stability, it was better to announce victory and ignore the losses.

On the other hand, industry was speaking to Congress, the media, and the public. Their message was that they—and the economy—had suffered a terrible blow and the selfish preservationists were locking up the National Forests. Even though industry won, they wisely pretended to have lost so that next time around they could claim it was their turn to win because the greedy environmentalists had won the previous fights.

A similar instance occurred in the fall of 1989, when Senator Mark Hatfield of Oregon and other members of Congress from the Northwest, at the behest of the timber industry, pushed through a legislative rider to an appropriations bill that prevented citizen lawsuits against National Forest timber sales. The upshot of this legislation, passed without adequate hearings or public consideration, was that more old growth will be sold for clearcutting on the National Forests in 1990–91 than ever before in an equivalent time span. The Sierra Club again claimed "victory." While this claim was true in a narrow sense—their lobbying had made the bill less rank than originally proposed—in reality this legislation was the gravest single defeat ever dealt the conservation movement by the timber industry.

Efficient operation has become the main concern of environmental groups. Once the national boards of these groups were made up of the leading conservationists in the country. Today, candidates for the Sierra Club board of directors emphasize their commitment to smooth business operations. Their election statements downplay any green fire burning in their eyes.

Professional staff are frequently unfamiliar with the intellectual discussions going on in the movement. There is a widening gulf between the political technicians and the thinkers, philosophers, and visionaries. (In fairness, it must be acknowledged that part of this failing falls at the feet of the philosophers who often wallow in obtuse academic discussions far removed from the real world of wilderness with which they ostensibly deal.)

There is a growing breach between grassroots volunteer activists and professionals. A salaried elite has formed that elevates itself above the volunteers. This professional cadre talks to itself and is often both defensive and arrogant in its dealings with the conservation grassroots. It frequently inhibits dissent or even public discussion of strategy and priorities.

The above trends and manifestations are admittedly general. There are certainly exceptions to them. It is necessary in demonstrating a shift in perspective to generalize. Although I have concentrated on The Wilderness Society and the Sierra Club (because I know them best), the phenomenon is movement-wide; even state groups are feeling the pressure from the national environmental establishment to become more "professional." The 1985–86 power struggle within Friends of the Earth had all the earmarks of a confrontation between radical amateurs and resource professionals.

There are a number of outstanding individuals working for national and state conservation groups who are worthy heirs to the tradition of Marshall and Brower, but they often seem like back eddies in the "professional" current within the stream of conservation.

Professional conservationists are here to stay. To prevent professionalism from further weakening the environmental movement, leaders of these organizations should carefully study the inherent problems of professionalism, and devise built-in safeguards against them. I do not want to rid the movement of paid staff (I would even agree that more paid staff would be helpful to the cause); I simply want to see paid staff function as effectively as possible and without usurping the proper leadership role of citizen activists. Some easy steps that can be implemented almost immediately, with little cost, and with, I think, great effectiveness against the pitfalls of professionalism are outlined below:

■ All employees of conservation groups (including those in administrative and clerical positions) should be required to take two weeks of paid vacation a year *in the wilderness.* There should be no exceptions. These should be real wilderness trips including at least one of a week's duration. It would be very simple, for example, to encourage Sierra Club staff to go on national SC outings.

The argument will be immediately raised that conservation staffers are overworked and do not have time for wilderness vacations. This is hypocrisy. One of the major arguments given for Wilderness Areas in the modern era is their role as refuges for recreation, for escape from the stresses of urban civilization. If wilderness lobbyists say they don't need "the tonic of the wilderness," then they do not believe in the cause they are promoting. On a personal note, I've found that I can accomplish more work in less time by frequently visiting Wilderness Areas and recharging my enthusiasm.

Some have argued against this proposal thus: A lobbyist or scientist working on acid rain, for example, has no need to visit the wild.

My reply is that such a person would be a far more effective advocate for the control of smelter and power plant emissions after canoeing lakes and rivers in New York's Adirondacks, the Maine Woods, or eastern Canada, and seeing the impact of acid precipitation on natural ecosystems.

■ Each group should hold at least one staff meeting annually in a rustic setting (not a plush ski lodge) with at least an overnight wilderness excursion before or after. There should be joint meetings in such surroundings between the issues staffs of the Sierra Club, The Wilderness Society, the National Audubon Society, Defenders of Wildlife, and state environmental groups.

■ Sierra Club regional representatives should not be hired or supervised by the national administration of the Club. Instead, grants should be made by the national club to each Regional Conservation Committee or to chapters to hire and fund conservation staff persons. These regional reps and their assistants would be employees of the RCC (or chapter) and would report to the RCC (or chapter), although they would work closely with the national conservation staff.

■ The Sierra Club's national board of directors has become largely occupied with mundane business matters and has too little to do with conservation. This has left a vacuum, allowing a staff clique to monopolize strategy, priorities, and implementation of conservation work. Perhaps the national SC board could be divided into two committees, one to deal with conservation, the other with administrative matters. Somehow, the grassroots activists of the Club need a greater role in directing action and fundamental strategy.

■ Although there are some outstanding conservationists on The Wilderness Society's council, the council seems largely composed of people with money or access to money (which is not necessarily incompatible with being a wilderness fanatic). Although raising money is important, the council needs to spend more time on conservation strategy and day-to-day issues.

As with the Sierra Club board of directors, perhaps the TWS council could be divided into two committees, one overseeing conservation and the other concentrating on business matters and raising money.

■ State wilderness groups in the West (including the Southeast Alaska Conservation Council, the Oregon Natural Resources Council, the California Wilderness Coalition, the Nevada Outdoor Recreation Association, the Committee for Idaho's High Desert, the Southern Utah Wilderness Alliance, the Colorado Environmental Coalition, and the Arizona Wilderness Coalition) should form the Western Wilderness Coalition and hire a lobbyist to represent them in Washington, D.C., so that they are not controlled by, or dependent on, the big national groups. The board of directors of the Western Wilderness Coalition could consist of the heads of the individual groups and could directly supervise the D.C. lobbyist. Regional groups east of the Rockies, such as the Texas Committee on Natural Resources and Friends of the Boundary Waters, should similarly organize. These coalitions should rent a house in Washington, D.C., where grassroots conservationists could stay when visiting the capital for lobbying. The Sierra Club and the National Audubon Society should also rent or buy a house in which their grassroots activists could stay when lobbying in D.C.

■ In recent years, a new kind of statewide or regional conservation group has emerged: organizations that operate within legal channels, but have more ambitious goals than do pragmatic mainstream groups. They propose biological corridors to connect core Wilderness Areas, wilderness recovery areas, and the reintroduction of extirpated species. These outfits, like the Alliance for the Wild Rockies, Native Forest Council, Klamath Forest Alliance, Greater [Cascades] Ecosystem Alliance, Virginians for Wilderness, Preserve Appalachian Wilderness, and Public Lands Action Network, should form a loose federation to better coordinate the development of ecological wilderness proposals and uncompromising stands on land management issues. Such a federation could move the Sierra Club and other mainstream organizations to stronger positions and action.

■ The Gang of Ten and other large environmental groups should jointly set up a fund to disburse at least a million dollars a year in $15,000 grants to individual conservation activists. There should be no strings attached to these "mini-MacArthur" grants, and the recipients should be free to work on whatever they wish. Since those chosen would be experienced, capable, self-starting activists, there should be no need for supervision or administrative overhead.

■ A ceiling should be placed on salaries paid by conservation groups. This would prevent many career-hopping political technicians from applying for such jobs, and would leave positions in the hands of those truly committed to preservation of natural diversity.

This would have the added advantage of allowing groups to hire more staff for the same cost. In 1979 the regional representatives of The Wilderness Society offered to take a cut in salary in order to hire more reps. We believed additional regional staff was needed to do the job. The executive director of TWS was aghast at our lack of "professionalism" and proceeded to fire much of the staff. (I was not fired. To my undying uneasiness, the guy liked me.)

■ A primary requisite for employment by a conservation group should be experience as a volunteer grassroots activist. Full-time paid staff positions should be earned by an apprenticeship as a grassroots conservation activist.

■ Staff members of conservation groups should be encouraged to be knowledgeable about the ongoing biological and philosophical discussions within the environmental movement through reading the important books and articles. Conservation groups could buy copies of the important works wholesale, and distribute them to staff. Seminal magazine articles should be distributed as well.

Again, there will be the objection that conservation staffers are overworked and do not have time to read. Medical professionals are also overworked, but they have to find time to read professional material, or they fall behind in their profession. It is just as important for a conservation professional to be current in the fields of conservation biology and environmental ethics.

■ Lobbyists for conservation groups in D.C. should be required to make at least four trips to the field annually. They should attend important conferences, meet with grassroots leaders, and visit key areas where national issues will emerge. For example, lobbyists working on wilderness should set foot in a few of the proposed areas in each state.

It has been argued that one does not need to know a particular area to fight effectively for it. My experience of eight years with The Wilderness Society clearly demonstrated to me that if I did not personally know an area, I would not fight as hard for it as for one I knew, that I would compromise on the area I didn't know. This is simply human nature. A lobbyist may not go the extra mile required to save an area if she has never seen it, if the Elk and snowstorms and bear tracks do not exist for her, if it is merely a disembodied name in a list.

■ Staff members should be encouraged to become personally involved in local conservation groups (for example, Sierra Club staff in D.C. should be involved in the Potomac Chapter) and in at least one local issue.

■ The Sierra Club/Wilderness Society Biennial Wilderness Conference should be revived as a forum for all wilderness preservation groups. Controversial speakers should be invited. Serious discussion, strategizing, speculating, and networking should take place between staff, volunteers, and academics. Dissent should not be stifled but rather encouraged.

■ Grassroots conservationists should develop a greater militancy to control the movement, a willingness to demand the replacement of professional staff who become aloof, co-opted, or elitist, and the confidence to stand up for decentralized decision-making.

There are other measures that could be taken to maximize the effectiveness of professional staff members; these are my suggestions to start the process. Acknowledging the inherent problem of bureaucracy in human organizations is the first step to developing organizations "with a human face." Addressing this typical problem is not a condemnation of those dedicated people who work for conservation groups (generally at lower salaries than they could receive elsewhere), or a claim that full-time, paid workers are not needed. It is, instead, an effort to maximize the effectiveness of the preservation movement and to safeguard it from co-option by the political establishment and from being weakened by the inherent nature of corporate bureaucracy.

We need professional lobbyists, scientists, attorneys, and accountants playing the game. But they need to up the ante and go on the offensive. We need competent administrators to manage multimillion-dollar-a-year budgets and large staffs. But they must be guided by the vision of Muir, Edge, Marshall, and Leopold—not by that of the Harvard Business School.

19 | Whither Earth First!?

A militant minority of wilderness-minded citizens must be on watch throughout the nation and vigilantly available for action.

—Aldo Leopold

In the fall of 1981, the *Albuquerque Journal* ran a front-page story on the budding Earth First! movement. Two mornings later I was awakened by a phone call. It was United States Senator Pete V. Domenici of New Mexico, calling from Washington, D.C.

"Dave! What's this bullshit I read in the *Journal*?"

It was a question I found difficult to answer.

I am a reluctant radical. I have always been a little embarrassed carrying a sign at a demonstration. I have never felt entirely comfortable sitting in front of a bulldozer or chaining myself to the doors of a National Park visitor center. I regret the need to damage private property of any sort. I do not like to take an uncompromising stand; I much prefer to sit down and work out differences respectfully and rationally. I feel more at home in the guise of a peacemaker than that of a rabble-rouser.

Earth First! was not conceived in an attitude of revolution for the hell of it, radicalism for its own sake, or antiestablishment rebellion. Earth First! was born in sorrow, in the hesitant realization by experienced but frustrated wilderness activists that polite, business-as-usual methods were not sufficient to save Aldo Leopold's "things natural, wild and free" from human greed.[1]

My original companions in Earth First! and I are radicals because the times demand it. Like Dr. Benjamin Spock during the Vietnam War, we are called to act because of a great injustice. Like Thomas Jefferson, George Washington, and John Adams, we pledge our lives, our fortunes, and our sacred honor because we are in the midst of crisis. Indeed, the biological crisis we face is no less demanding than the political crisis they faced. Like the Reverend Martin Luther King, Jr., we break the law because the law serves oppression.

We do this with reluctance, and with sorrow. We do it recognizing once again with Tom Paine that "these are the times that try men's souls."

We do not engage in radical action because we are primarily motivated by opposition to authority, because we are antinomians, but because we are *for* something—the beauty, wisdom, and abundance of this living planet.

After ten years, Earth First! has accomplished much of what it initially set out to do (although far more remains to be done). While our efforts have resulted in actually stopping several bad projects, such as the Bald Mountain logging road in Oregon's Siskiyou National Forest, most of our success has been in influencing and strengthening the environmental movement. In 1980, Earth First! became the first conservation group to demand a complete halt to the logging of old growth on the National Forests. The dogged

---

1. Despite this disinclination to act in a radical fashion, there was a bedrock understanding that the biocentric implication of wilderness preservation—that Earth was not for human beings alone—was a thoroughly radical concept at odds with the clanking camshaft of civilization. We founders of Earth First! weren't uncommonly clever to realize this; the revolutionary significance of the wilderness idea has long been apparent to conservationists. Many women and men who thrill after the call of the wild have realized that they do not stand simply off to the side of the rush hour of progress, but squarely in its path. Of course, many have also argued for wilderness preservation from the anthropocentric view that wilderness has benefits for human beings—recreation, science, contemplation, aesthetics, watershed protection, and so forth.

determination and courage of activists in the Pacific Northwest, the Northern Rockies, and the Southwest has not only slowed the destruction of ancient forests, but has thrust the issue into the national spotlight and persuaded mainstream groups like the National Audubon Society and The Wilderness Society to make ancient-forest preservation a priority. Earth First!–inspired demonstrations against Burger King and the World Bank have helped boost the preservation of tropical rain forests to international prominence. Since 1980, Earth First! has led the effort to reframe the question of wilderness preservation from an aesthetic and utilitarian one to an ecological one, from a focus on scenery and recreation to a focus on biological diversity.

Similarly, we have gone beyond the limited agenda of mainstream conservation groups to protect *a portion* of the remaining wilderness by calling for the reintroduction of extirpated species and the restoration of vast wilderness tracts. We have brought the discussion of biocentric philosophy—Deep Ecology—out of dusty academic journals. We have effectively introduced nonviolent civil disobedience into the repertoire of wildland preservation activism. We have also helped to jolt the conservation movement out of its middle-age lethargy and reinspire it with passion, joy, and humor. In doing all of this, Earth First! has restructured the conservation spectrum and redefined the parameters of debate on ecological matters.

Of course, no one can stir the pot like this without attracting headlines. As a result of Earth First!'s notoriety, everyone seems to want to change us. From one side there are concerted efforts to moderate us, mellow us out, and sanitize our views; from another side have come efforts to make us radical in a traditional leftist sense; and there are ongoing efforts by the powers that be to wipe us out entirely.

As was to be expected, much of the pressure to change Earth First! has come from the political establishment and from mainstream conservation groups. This is typical of our system: Radicals are effectively dealt with by giving them a place in the structure, where they are co-opted.

Surprisingly, however, a more vehement spurning of wilderness fundamentalism comes from those who wear their "radicalness" as a badge. From the beginning, some people have been attracted to Earth First! because it represented to them a reincarnation of the style and intensity of the New Left. Early on, we received letters chiding us any time we deviated from a leftist, "politically correct" line, and urging us to develop "a more sophisticated critique of capitalism." The question of overpopulation is a particular bugbear with the left, and any article in the *Earth First! Journal* discussing that snarled issue has been destined to collect an immediate, albeit small, volley of darts. Coincidental with this growing denunciation from the left was the increasing involvement in EF! of generally young men and women with vaguely held "anarchist" ideas and life-styles.

Diversity in abilities, life-styles, talents, personalities, and even ideas partially accounts for what Earth First! has accomplished. Just as a diverse ecosystem is more stable, a diverse social movement can be stronger. Nonetheless, while diversity can strengthen and stabilize a group, too much diversity can fracture and immobilize it. The increased visibility of Earth First! has made it an "in" group, thereby attracting to the excitement and notoriety some people who do not understand what biocentrism really means. As a result, Earth First! has become disparate, and there are disagreements over matters of philosophy and style that threaten to compromise our basic tenets or make us impotent.

As originally brewed, Earth First! was not part of the left or of the Rainbow Tribe (the major venue these days for hippies), or a reincarnation of Abbie Hoffman's Yippies.[2] We were not part of the animal rights, anarchist, peace, social justice, antinuclear, nonviolent, neo-pagan, native rights, or Green movements. Over the years

2. There was, however, a certain Yippie flavor to the early Earth First! Mike Roselle, one of the EF! founders, had been a Yippie activist as a teenager in the early 1970s. During the first couple of years of EF!, as we made our break from the conservation mainstream, there was an emphasis on zany and colorful actions. Despite this friskiness, our influence from the Yippies was that of technique, not philosophy.

we have had varying degrees of affinity with some of these, but we have not been in any way subject to them. We have been the Earth First! movement. As such, we have not been the entire radical environmental movement or the entire Deep Ecology movement. There is plenty of elbow room in the conservation spectrum on the rambunctious side of the Sierra Club, far more than Greenpeace, Earth Island Institute, Rainforest Action Network, Sea Shepherd, and Earth First! can fill. Had Earth First! tried to straddle too much of the "radical environmental movement" we would have fallen flat on our face.

It is absolutely essential to understand that Earth First! did not emerge from the anarchist movement, or from the left. Neither were we born of sea foam, like *The Birth of Venus*. Earth First! came directly out of the public lands conservation movement—out of The Wilderness Society, the Sierra Club, Friends of the Earth, and the National Audubon Society. Public lands issues and wilderness have been central to us from our formation. Individuals have certainly come to Earth First! from other movements, some EF!ers have primary connections to other movements, and EF! cooperates with other movements on certain issues; but they were not our source and have not been our primary thrust.

In recent years, however, virtually every large Earth First! gathering has spawned arguments about style and substance. More and more old-time EF!ers fail to attend the annual Round River Rendezvous because they feel it has transmogrified into a hippie/punk revel. Some leading Earth First! activists have turned from defending the wilderness to storming the barricades of capitalism. In Washington and New Mexico, conservation-oriented EF!ers, embarrassed by revolution-for-the-hell-of-it rowdies, have regrouped under other names. The anarchist faction, frustrated with what they see as a stodgy *Earth First! Journal,* is now producing a "punkzine" that aims to cultivate an impalpable "wildness within." Even their more sober comrades in direct action criticize the *Earth First! Journal* for wasting space on articles about conservation biology and biocentric philosophy when that space could be used for more news pieces and photographs about *their* actions. Many Earth First!ers have begun to wonder if the movement will continue to hold together under that name.

I'm not sure it matters.

Thomas Jefferson suggested that a revolution every generation may be desirable to prevent bureaucracy and stagnation.

In order to understand what is happening to Earth First!, let us go to evolutionary ecology and consider a generalist species in a new habitat with many available niches—say, a finch blown over to the Galápagos Islands from South America. Slowly, different populations of that generalist species adapt to more effectively exploit the different niches and evolve into separate species. Oftentimes, external environmental stresses push a generalist species toward faster differentiation into separate, specifically adapted sister species.

This is what has happened to Earth First! From the beginning, the Earth First! movement has had three major strains: monkeywrenching; biocentrism and ecological wilderness preservation/restoration; and confrontational direct action, both legal (demonstrations) and illegal (civil disobedience). Different personalities have been attracted to Earth First! by each of these strains. Those given to better exploiting the monkeywrenching, direct action, and conservation biology niches have lately been diverging. Recent pressure and infiltration from the class-struggle/social-justice left and predation from the FBI have hastened this divergence. Unless we can adapt to both this changed environment and this divergence within our gene pool, we will become extinct.

One would-be leader of a "new" Earth First! has said that it is the nature of movements to *move.* He is correct. In recent years, a strong faction has risen on the West Coast that finds its place more with the left than with the conservation movement and is inspired more by the writings of Abbie Hoffman than those of Edward Abbey. That is fine. If the Earth First! movement is to become a revival of the New Left with an environmental focus, so be it. Such a group of shock troops is needed.

The problem, however, is when excessive internal debate about style, strategy, and substance leads to infighting that keeps all of us from the real job—fighting the vandals looting the riches of this Earth. For me, I stand by the biocentric guidelines I offered at the Grand Canyon in 1987 and find it difficult to advocate ecological wilderness restoration within a movement now dominated by anti-capitalist rhetoric and an overwhelming emphasis on direct action to the exclusion of other traditional Earth First! techniques. I am not an anarchist or a Yippie. I am a conservationist. I believe that human overpopulation is the fundamental problem on Earth today.[3] Although I will continue to applaud the courageous actions of those operating with the Earth First! name, it is time for me to build a campfire elsewhere.

In other words, I am no longer part of the Earth First! movement. I no longer represent it and I am no longer represented by it.

There are a variety of scenarios of what may happen to Earth First! under the triple pressures of external harassment from the FBI, encouragement to moderate from the establishment, and internal divisiveness. What I see happening now to the Earth First! movement is what happened to the Greens in West Germany—a concerted effort to transform an ecological group into a leftist group. I also see a transformation to a wholeheartedly counterculture/antiestablishment style and the abandonment of biocentrism in favor of humanism.

The most likely scenario in my view is that those most interested in conservation biology and big wilderness, myself included, will promote their ideas under a name other than Earth First! as that name becomes muddied by haphazard vandalism, kiddie stunts like flag burnings and shopping mall puke-ins, Yippie-style actions, and

3. William R. Catton, Jr., restates Malthus's dictum in ecological terms as: "The cumulative biotic potential of any species exceeds the carrying capacity of its habitat." This is exactly the case with human beings on this planet today. I do not think that believing so means one is racist, fascist, imperialist, sexist, or misanthropic, even if it is politically incorrect for cornucopians of the left, right, and middle. Catton's book *Overshoot: The Ecological Basis of Revolutionary Change* (Urbana, Ill., and Chicago: University of Illinois Press, 1982) is, I firmly believe, the most important book published during the last forty years. It is must reading for anyone who wishes to understand or, more importantly, do something about the ecological crisis we humans have created on this planet.

class-struggle slogans. Wise monkeywrenchers can more safely practice their midnight art by not drawing attention to themselves as known activists with any group, especially Earth First! Those committed to confrontational civil disobedience may find their position more secure by eschewing the advocacy of property destruction. Earth First! may pass away as a particular focus, or it may become more and more dominated by direct action advocates from the anticapitalist left.

Regardless of what happens, the work will go on.

This is the beauty of Earth First!, and the reason it cannot be stamped out even if it ceases to exist as a distinct entity. There will continue to be gutsy, never-say-die tree climbers and bulldozer blockaders; there will continue to be biocentric, activist biologists working for other species; there will be increasing numbers of stronger, bolder, less compromising mainstream activists; and there will be even more anonymous ecodefenders messing around with big yellow machines in the dark of the new moon.

No matter what happens, the hot green chiles from the original Earth First! will remain a permanent part of the conservation movement. Neither feral adolescents, the co-opting whirlpool of the American mainstream, nor even the Federal Bureau of Investigation can stop us.

Acknowledgments

I have been fortunate in my twenty years of conservation work to have been shown the way by some of the finest conservationists of the generation and a half before mine. My mentors include Stewart Brandborg, David Brower, Brant Calkin, Harry Crandell, Ernie Dickerman, Michael Frome, Celia Hunter, Martin Litton, Jerry Mallett, John McComb, Clif Merritt, Mardie Murie, Michael Nadel, Jim Stowe, Charlie Watson, and Art Wright. I thank them for their vision and good works, and for pointing me down the trail.

Just as no man is an island, no one's ideas are his alone. What is offered in this book comes from countless discussions around the campfire, over beers and cigars, while gnawing pork-chop bones, and sweating up desert peaks. Of course, only I bear responsibility for what is in this book, but what is here was roused by my encounters with these friends.

Five in particular stand out. They are Bart Koehler, Howie Wolke, Ron Kezar, John Davis, and my wife, Nancy Morton. I can't begin to express my thanks for their companionship over the years. Howie, Ron, and Nancy were also invaluable critics in the creation of this book; John Davis even more so. John not only parried back and forth with me on the ideas herein, but made important editorial suggestions through several drafts. My reliance on John as a critic and editor is shown by my reluctance to write anything without his review, although we are a continent apart.

Other friends who reviewed and otherwise specifically assisted with parts of this book are Ernie Dickerman, Jim Eaton, Mitch Friedman, Denzel and Nancy Ferguson, Celia Hunter, Lynn Jacobs, Clif Merritt, Rod Mondt, Reed Noss, Lester Rhodes, Gerry Spence, Dale Turner, Charlie Watson, George Wuerthner, and Nancy Zierenberg. Thanks, friends.

There are many others to thank for helping me hone my ideas over the last twenty years. Among these friends are Edward and Clarke Abbey, Ric Bailey, Wendell Berry, Joe Bernhard, Bill Bishop, Ed Caldwell, Jasper Carlton, Dan Conner, Bill Devall, Barbara Dugelby, Jack Dykinga, Jeff Elliott, Steve Evans, Roger Featherstone, Randy Hayes, Bruce Hayse, Paul Hirt, David Johns, Huey Johnson, Kay Johnson, Bob Kaspar, Andy Kerr, Bob Langsenkamp, Jack Loeffler, Wes Leonard, Tim Mahoney, Christoph Manes, Stephanie Mills, Susan Morgan, Bob Mueller, Arne Naess, Shaaron Netherton, Doug and Lisa Peacock, Mike Roselle, Ken Sanders, Jamie Sayen, Roger Scholl, Debbie Sease, John Seed, George Sessions, Ken Sleight, Gary Snyder, Kris and Les Sommerville, Karen Tanner, Paul Watson, and Marcy Willow. In addition to these, I have been motivated to carry on by the example set by hundreds of grassroots defenders of the planet.

Among the many writers and reporters who have interviewed me, several stand out for their thoughtful, provocative questions, which prodded me to think more deeply about my answers. They are Chuck Bowden, Bill Kittredge, Marley Klaus, Michael Lerner, David Quammen, Rik Scarce, and Susan Zakin.

Often a song can say more than can prose or talk. The songwriters and singers who have especially inspired me are Greg Keeler, Katie Lee, Dana Lyons, Bill Oliver, Cecelia Ostrow, Johnny Sagebrush, Walkin' Jim Stoltz, and Glen Waldeck.

ACKNOWLEDGMENTS

In a literate society like ours, one engages in a wide-ranging dialogue through time and space via books. The authors who have most sparked my own thinking are Edward Abbey, Morris Berman, Wendell Berry, Murray Bookchin, Rachel Carson, Vernon Gill Carter and Tom Dale, William Catton, Michael Cohen, Alfred Crosby, Bill Devall, Paul and Anne Ehrlich, David Ehrenfeld, Michael Frome, Romain Gary, Dolores LaChapelle, Aldo Leopold, Bob Marshall, Paul Martin, Chris Maser, Peter Matthiessen, William McNeill, John Muir, Lewis Mumford, Gary Nabhan, Roderick Nash, Tom Paine, Alfred Runte, Paul B. Sears, George Sessions, Paul Shepard, Gary Snyder, Michael Soulé, Henry David Thoreau, Frederick Turner, Donald Worster, and George Wuerthner.

My editor at Harmony Books, Michael Pietsch, did an excellent job of helping me turn a collection of essays into a book. It's been a pleasure to work with him as it has been with my agent, Timothy Schaffner.

I'd also like to thank my mother (Lorane Foreman), sister (Roxanne Pacheco), nephew and nieces (Gerard and Monica Pacheco, Lisa Foreman) and father- and mother-in-law (Bob and Nell Morton), for putting up with kin like me.

Finally, I must thank all the people who have so selflessly contributed in so many ways to my legal defense, and especially to my attorneys Gerry Spence, Richard Sherman, Sam Guiberson, and Roberta Sealey, John Ackerman, Cat Bennett, Kent Spence, and Ed Moriarity.

Index

225

INDEX

INDEX

INDEX